DEFORMATION MECHANISMS, TEXTURE, AND ANISOTROPY IN ZIRCONIUM AND ZIRCALOY

Erich Tenckhoff

ASTM

SPECIAL TECHNICAL PUBLICATION (STP 966)
1916 Race Street, Philadelphia, PA 19103

Library of Congress Cataloging-in-Publication Data

Tenckhoff, Erich.
[Verformungsmechanismen, Textur und Anisotropie in Zirkonium und Zircaloy. English]
Deformation mechanisms, texture, and anisotropy in zirconium and Zircaloy/Erich Tenckhoff.
(ASTM special technical publication; STP 966)
Translation of: Verformungsmechanismen, Textur und Anisotropie in Zirkonium und Zircaloy.
"(STP)04-966000-35."
Bibliography: p.
Includes index.
ISBN 0-8031-0958-x
1. Zirconium. 2. Zirconium alloys. 3. Deformations (Mechanics) 4. Texture (Metallography) 5. Anisotropy. I. Title. II. Series: ASTM special technical publication; 966.
TN693.Z5T4613 1988
620.1′8935—dc19

NOTE

The Society is not responsible, as a body, for the statements and opinions advanced in this publication.
This work was originally published by Gebrüder Borntraeger Verlagsbuchhandlung, Berlin, F.R. Germany, as No. 5 of the Materialkundlich Technische Reihe in 1978.

Printed in Philadelphia, PA
March 1988

Preface

This book is aimed at students and practicing engineers who wish to extend their knowledge of the effects of deformation mechanisms on both texture formation and mechanical anisotropy. Following a general review of the deformation mechanisms in hexagonal close packed (hcp) metals, the examples of zirconium and Zircaloy (the latter being of technical importance for light water reactor fuel element cladding tubes) are used to illustrate the interactions involved. A clarification of the relationships is of interest to theoreticians, because it contributes to understanding the theory of deformation during texture formation in hcp metals. By allowing for these relationships, it is possible for the practicing engineer to select the texture of zirconium and Zircaloy semifinished products by choosing the appropriate deformation parameters, so they can then be optimally adapted to the operational demands. The knowledge gained can be applied similarly to other hcp metals, if one allows for the metal specific perimeters of the hexagonal structure.

Acknowledgments

The author would like to thank Mr. P. L. Rittenhouse, Dr. G. R. Love, and Dr. R. O. Williams for providing him with stimulating suggestions in the course of numerous technical discussions during his work at Oak Ridge National Laboratories. Professor B. Ilschner, Erlangen, F.R. Germany, is owed particular thanks for his continual support and encouraging interest in the continuation of this work. Professor M. von Heimendahl, Erlangen, and Dr. K.-H. Matucha, Frankfurt, F.R. Germany, are also thanked for their stimulating technical discussions. Finally, the author would like to thank Professor I. Grewen, Bonn, F.R. Germany, and Professor U. Rösler and Professor K. Zwicker, Erlangen, for their valuable suggestions.

Contents

Chapter 1—Introduction

The deformation systems in hexagonal close-packed (hcp) metals are not as symmetrically distributed as in cubic metals. Furthermore, because the primary slip systems are not as numerous and are limited to deformations in the *a* direction, twinning competes with slip in plastic deformation and can, depending on the deformation conditions, play an essential role [*1–4*].[1]

In hcp metals, the low offer of slip systems, their asymmetrical distribution, and the strict crystallographic orientation relationships for twinning result in the formation of a strong deformation texture. (If the material is subsequently heat-treated, a pronounced annealing texture develops) [*5–7*].

For textured materials, on the other hand, the deformation mechanisms are also responsible for the strong anisotropy of the mechanical properties [*8–10*].

The foregoing points represent three aspects that are mutually interdependent. This paper is correspondingly divided into three sections: by way of introduction, the generally established relationships for deformation mechanisms in hcp metals are given, together with a discussion of their dependence on the metal-specific parameters, such as the c/a axial ratio, the transformation behavior, and the correlated stacking fault energy. This includes a detailed consideration of deformation mechanisms in zirconium and Zircaloy. In the second and third section, the influences of the deformation mechanisms on the texture development and on the mechanical anisotropy are discussed. These interactions can be transferred similarly to other hcp metals, if one allows for the metal-specific parameters of the hexagonal structure.

The hcp metals are particularly interesting in the investigation of the preceding relationships for the following reasons:

1. Owing to their marked structural anisotropy, the effects under discussion and their interactions become particularly evident, even though

[1] The italic numbers in brackets refer to the list of references appended to this book.

the relationships here are much more complicated than with cubic metals, for example.

2. They have not as yet been as thoroughly examined.
3. Their technical application is gaining in importance [*11*].

For example, zirconium and zirconium-rich alloys, to which Zircaloy belongs, are employed in nuclear reactors, specifically for the fuel element cladding tubes of light-water reactors, due to their small capture cross-section for thermal neutrons, their relatively good high-temperature strength, and their good resistance to corrosion.

Titanium, which resembles zirconium in its chemical and physical properties, together with titanium-rich alloys are used for chemical plant construction as well as in aerospace technology.

Magnesium and its alloys are employed as structural materials in aviation technology because of their high strength-to-weight ratio. They are also used in pyro- and galvanotechnics because of their strong chemical activity.

Zinc and zinc alloys are used for protective coatings and chemical plant construction. Zinc is also used as an alloy component in brasses and in aluminum-zinc-magnesium (AlZnMg) alloys.

Beryllium is used in nuclear and X-ray measurement technology due to its low absorption coefficient for thermal neutrons and X-rays. It is also used in aerospace technology due to its high strength-to-weight ratio.

In addition, elements such as cadmium and cobalt are used specifically as alloy components in wear-resistant and bearing materials as well as in creep-resistant materials.

Chapter 2—Deformation Mechanisms

2.1 Deformation Mechanisms in HCP Metals

The deformation mechanisms of hcp metals are more complex and less well investigated than those of, for example, face-centered cubic (fcc) and body-centered cubic (bcc) metals [*4,12–15*].

Although hcp metals normally form what is considered a single class, the individual metals differ in their crystallographic structure. The c/a axial ratio varies from one metal to the other and can attain a value larger or smaller than the ideal sphere packing (Fig. 1).

The angular relationships between the corresponding crystallographic planes are thereby altered. Independent of the absolute interplanar spacing, which varies from metal to metal regardless of its structure, this purely steric effect of the c/a axial ratio is an initial, important aspect that prevents the hcp metals from being considered en bloc. This is in contrast to cubic metals, for which the angular relationships between the corresponding planes are identical.

Owing to the low offer of possible slip systems within one slip system family and to its asymmetrical distribution over the reference sphere, various deformation systems can become operative. Various primary and secondary slip and twinning systems are activated, which, in addition, can exhibit various critical resolved shear stresses. In contrast to this, the relationships for fcc metals are considerably simpler, whereby 12 possible slip systems belonging to the same primary slip system $\{111\}\ \langle 1\bar{1}0\rangle$ are distributed symmetrically over the whole reference sphere. Therefore, in fcc metals usually only slip systems of this one family become operative; under uniaxial loading and at low deformation rates, single slip occurs, whereas multiaxial loading as well as higher deformation rates lead to duplex or multiple slip of mutually equivalent systems. The following sections of this chapter discuss the extent to which the relationship of hcp metals are more complicated.

2.1.1 Slip Modes

There are three well-established laws governing the slip behavior of metals, namely:

(*a*) From a given set of slip planes and directions, the crystal operates on that system (plane and direction) for which the difference between the resolved shear stress and the critical resolved shear stress is largest.

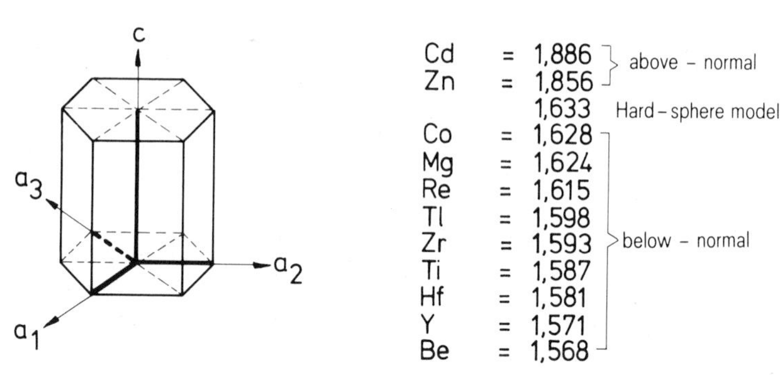

FIG. 1—*The* c/a *axial ratio for various hcp metals.*

(*b*) The slip plane is normally the plane whose interplanar spacing, that is, whose atomic density, is relatively the largest.

(*c*) The slip direction is nearly always the one that contains the shortest Burgers vector.

Law (*a*) corresponds to Schmid's law of resolved shear stress [*16*], initially found by experiment, that is discussed in more detail later in conjunction with the Schmid factor (Section 2.3.1). Laws (*b*) and (*c*) are included in the Peierls-Naborro model [*17,18*]. This model considers the undistorted lattice as a classical elastic continuum [*17*] and assumes forces for the atomic interactions along the dislocation plane that periodically vary corresponding to the sinus function of the relative atomic displacements [*18*]. It should be mentioned that the model only allows for estimations of order of magnitude; exact calculations must take into account the interactions in the distorted crystal lattice and the atomic structure, that is, the electron structure of the atoms, the bond energies, and the geometric arrangement. In spite of its inaccuracy, the Peierls-Naborro model permits the general trend for metallic and multipolar bonds to be ascertained, whereby the Peierls-Nabarro forces are small for short Burgers vectors in planes with high atomic density, that is, with large interplanar spacing. Sagel and Zwicker [*19*] have demonstrated that the model is also applicable to hcp metals using the example of titanium.

2.1.1.1 Slip Plane—Although the interplanar spacing of the basal plane is a constant $c/2$ for hcp metals, the other types of planes, the prism and pyramidal planes, each have an irregular interplanar spacing, as shown, for instance, in Fig. 2 for certain planes.

Taking into account the A-B stacking sequence, the clearly visible waviness of the non-basal planes and their irregular spacing explain why the slip models for prism and pyramidal planes are so complicated (Section 2.1.1.2).

2.1.1.1.1 Effect of the c/a *Axial Ratio*—The interplanar spacing and the packing density vary with the c/a axial ratio of the hexagonal unit cell. For above normal and ideal axial ratios, the basal plane (0001) is the one most densely packed. However, if the c/a axial ratio is less than $\sqrt{3}$, then the prism plane is on average more densely packed than the basal plane (see Fig. 2). On the basis of the Peierls-Nabarro model [*17–18*], one would therefore expect that, for example, during slipping in the a-direction, a change in slip plane from (0001) to $\{10\bar{1}0\}$ would take place (Fig. 3). Apart from certain exceptions, such as magnesium (Mg), cobalt (Co), and beryllium (Be), this concept is valid and can be applied analogously to other slip systems. An attempt is made to explain these exceptions with the aid of further factors influencing the choice of slip plane, that is, the stacking fault energy and the transformation behavior (Section 2.1.1.2).

2.1.1.2 Slip Direction—The possible Burgers vectors in hcp metals can be represented (Fig. 4) by means of a double tetrahedron, which is derived correspondingly from the Thompson (reference) tetrahedron for fcc metals [*20*]. There are thus three possible types of perfect dislocations and three possible types of partial dislocations.

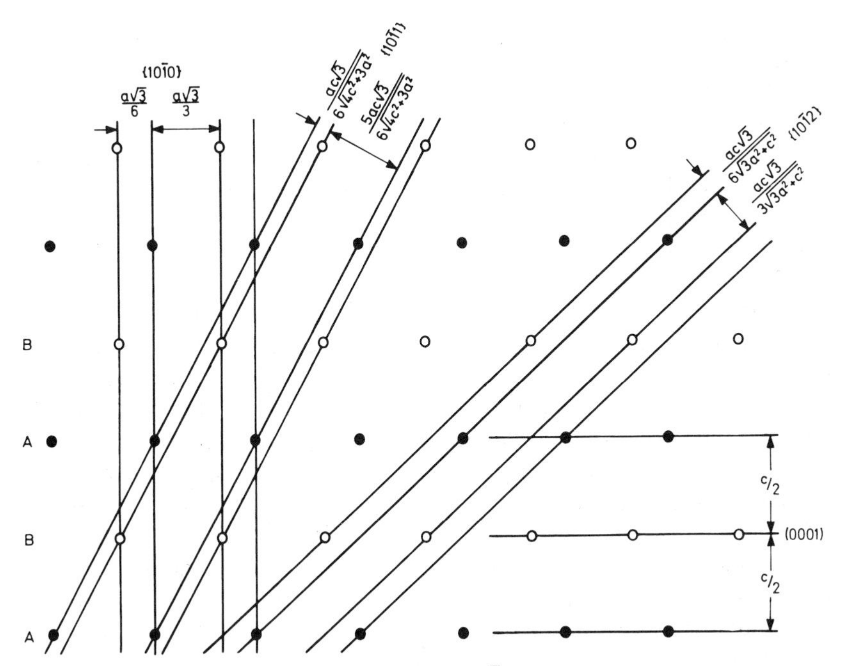

FIG. 2—*Projections of the atomic layers in the* $(1\bar{2}10)$ *plane (● atoms in Sequence A, ○ atoms in Sequence B). The lines show the traces and interplanar spacings of* (0001), $\{10\bar{1}2\}$, $\{10\bar{1}1\}$, *and* $\{10\bar{1}0\}$ *planes for ideal sphere packing* (c/a = *1.633*), *corresponding to Ref* 4.

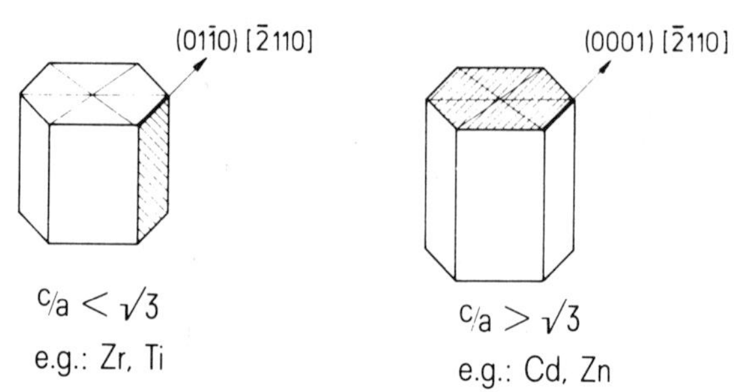

FIG. 3—*Dependence of the slip plane for* a *dislocations on the* c/a *axial ratio, corresponding to the Peierls-Naborro model.*

Types	
I	Six perfect *a* dislocations with the Burgers vectors AB, BC, CA, BA, CB, and AC of the 1/3 ⟨11$\bar{2}$0⟩ type.
II	Two perfect *c* dislocations with the Burgers vectors ST and TS of the ⟨0001⟩ type.
III	Twelve perfect (*c* + *a*) dislocations SA/TB, SA/TC, SB/TA, SB/TC, SC/TA, and SC/TB as well as their negative indices of the 1/3 ⟨11$\bar{2}$3⟩ type.
IV	Six partial *p* dislocations with the Burgers vectors Aσ, Bσ, Cσ, σA, σB, and σC of the 1/3 ⟨10$\bar{1}$0⟩ type.
V	Four partial *c*/2 dislocations with the Burgers vectors σS, σT, Sσ, and Tσ of the 1/2 ⟨0001⟩ type.
VI	Twelve partial (*c*/2 + *p*) dislocations with the Burgers vectors AS, BS, CS, AT, BT, and CT as well as their negative indices of the 1/6 ⟨20$\bar{2}$3⟩ type.

Assuming an ideal sphere packing ($c/a = 1.633$), the squares of the vector moduli are listed in Table 1, these being proportional at a first approximation to the elastic energies connected with the dislocations. In accordance with this consideration, those dislocations that have the shortest Burgers vector become effective for a given slip direction. Within the class of perfect dislocations, from this viewpoint *a* dislocations are favored over *c* dislocations and these, in turn, over ($c + a$) dislocations.

Because the Burgers vectors of the *a* and *c* dislocations are perpendicular to one another, they cannot mutually interact. On the other hand, the ($c + a$) dislocations can, with an energy gain, interact with *a* or *c* dislocations and form stable dislocation nodes, assuming that the relatively long ($c + a$) Burgers vectors can exist on the basis of the preceding energy criterion. Corresponding energy considerations can also be made for the partial dislocations, which are energetically more favorable owing to their

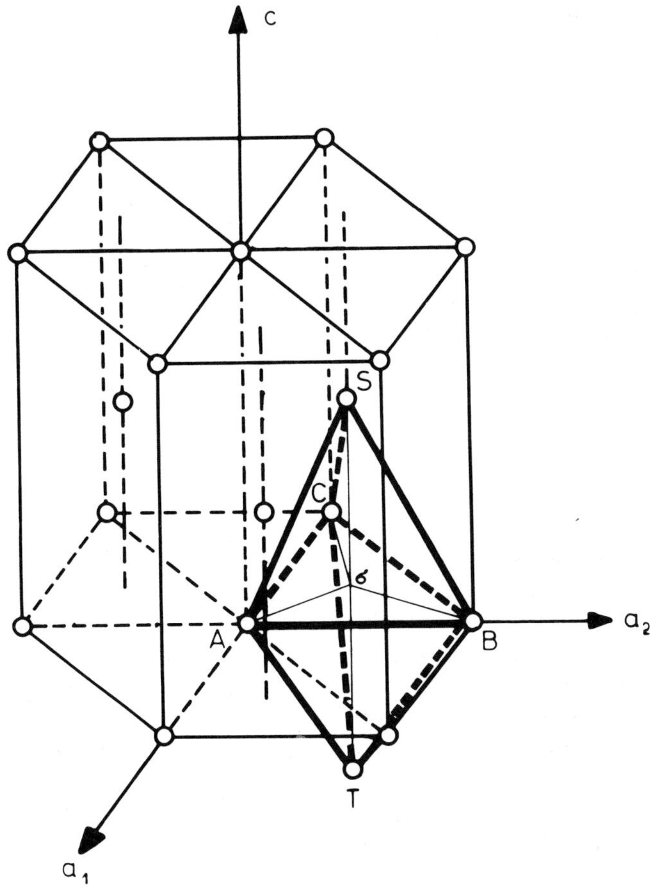

FIG. 4—*Theoretically possible Burgers vectors in an hcp lattice (according to Ref 21).*

TABLE 1—*Vector energies for the possible dislocation types in hcp metals for an ideal* c/a *axial ratio.*

	Dislocation Type					
	I	II	III	IV	V	VI
Designation	a	c	$(c + a)$	p	$c/2$	$\left(\frac{c}{2} + p\right)$
Number of dislocations per type	6	2	12	6	4	12
Burgers vector	1/3 ⟨11$\bar{2}$0⟩	⟨0001⟩	1/3 ⟨11$\bar{2}$3⟩	1/3 ⟨10$\bar{1}$0⟩	1/2 ⟨0001⟩	1/6 ⟨20$\bar{2}$3⟩
Vector energy	a^2	$c^2 = 8/3\ a^2$	$11/3\ a^2$	$1/3\ a^2$	$2/3\ a^2$	a^2

shorter Burgers vectors. However, if and to what extent a dissociation into two dislocation partials is possible among others, also depends on the stacking fault energy.

2.1.1.2.1 Influence of the Stacking Fault Energy, γ

(*a*) Dissociation of *a* Dislocations in the Basal Plane

For low stacking fault energy, γ, the *a* dislocation in the basal plane can be dissociated with energy gain into two Shockley partials, *p*, surrounding an intrinsic stacking fault that violates two next-nearest neighbors in the stacking sequence [*13–14*] (see Fig. 4).

$$AB \rightarrow A\sigma + \sigma B \tag{1}$$

$$\frac{1}{3}[11\bar{2}0] \rightarrow \frac{1}{3}[10\bar{1}0] + \frac{1}{3}[01\bar{1}0] \tag{2}$$

For this dislocation reaction in the basal plane, which occurs within a thin intermediate layer in a likewise densely packed fcc stacking sequence, there are two possible slip sequences (Fig. 5): either a B layer slips over an A layer (that is, σB follows σA as in Fig. 5*a*), or an A layer slips over a B layer (that is, Aσ follows Bσ as in Fig. 5*b*). The ribbon width is inversely proportional to the stacking fault energy, γ, that is either known or estimated for *a* dislocations in basal planes for several hcp metals (Table 2).

The *a* dislocation dissociated into 2*p* partial dislocations is confined

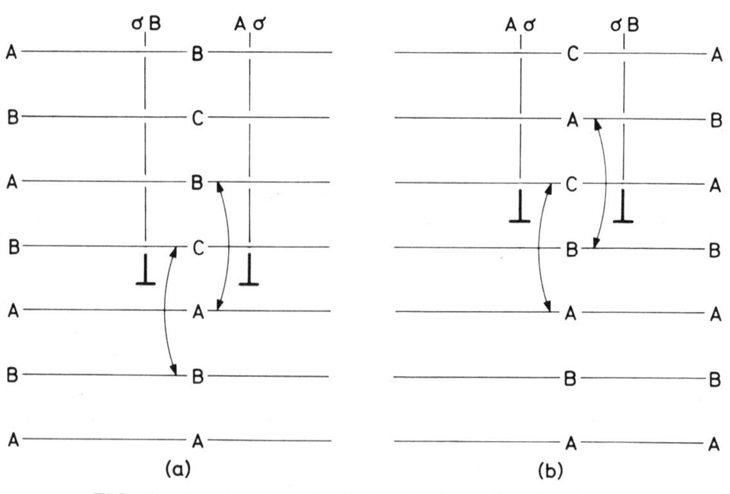

FIG. 5—*Stacking faults in the basal plane of an hcp lattice.*

TABLE 2—*Stacking fault energy, γ, for* a *dislocations in the basal plane*[a] *respectively in the prism plane*[b] *of various hcp metals.*

Material	Stacking Fault Energy, γ J/m² × 10⁻³ ≙ erg/cm²	Ref
Cd	250 to 300[a]	*13*
	170[a]	*15*
	150[a]	*22*
Zn	250 to 300[a]	*13*
	300[a]	*15, 22, 23, 24*
Co	25[a]	*13*
	26[a]	*15, 23*
Mg	250 to 300[a]	*13*
	300[a]	*15, 24*
	60[a]	*25*
	10 to 21[a] (calculated)	*26*
Zr	250 to 300[a]	*13*
	56[b]	*28*
Ti	250 to 300[a]	*13*
	300[a]	*15, 27*
	145[b]	*29*
Be	250 to 300[a]	*13*
	~180[a]	*15, 27*
	>190[a]	*25*
	279 to 1147[a] (calculated)	*26*
	≧790[a] (anticipated)	*30*
	≧1190[b] (anticipated)	*30*

solely to slip in the basal plane. The coalescence of screw components can cause cross-slip in first-order prism and pyramidal planes of the $\{10\bar{1}n\}$ type, although each of the given $\{10\bar{1}n\}$ planes can only contain one of the three possible $\pm 1/3\ \langle 11\bar{2}0\rangle$ Burgers vectors. A dislocation with a first-order Burgers vector, however, cannot slip in second-order prism and pyramidal planes of the $\{11\bar{2}n\}$ type.

The dissociation of *a* dislocations into partial dislocations in the basal plane could also be at least indirectly confirmed by experiment (Table 2). Thus, for cobalt, which has a low stacking fault energy, corresponding dislocation reactions were observed in the basal plane [*15*]. No observations of intrinsic stacking faults could be made with other metals having a high stacking fault energy, obviously because the dislocations are not dissociated sufficiently.

(*b*) Dissociation of *a* Dislocations in First-Order Prism Planes $\{10\bar{1}0\}$

A dissociation of *a* dislocations in the first-order prism plane $\{10\bar{1}0\}$ is possible, according to Refs *31* and *32*

$$\frac{1}{3}[1\bar{2}10] \rightarrow \frac{1}{18}[4\bar{6}23] + \frac{1}{18}[2\bar{6}43] \qquad (3)$$

or, according to Ref *33*

$$\frac{1}{3}[1\bar{2}10] \rightarrow \frac{1}{6}[1\bar{2}1\bar{1}] + \frac{1}{6}[1\bar{2}11] \quad (4)$$

If one takes into account the waviness of the prism planes and their irregular spacing from one another, the dissociation of the *a* dislocation in the prism plane according to Eq 3 is energetically more favorable [*4*] compared with Eq 4; both reactions, however, should be energetically inferior compared with the dissociation in the basal plane. This could explain why even for these metals with a slightly below-normal c/a axial ratio (for example, cobalt and magnesium), basal slip is favored over prism slip.

The question as to why beryllium, which constitutes an exception by having the lowest c/a axial ratio, slips in the *a* direction not only in the prism plane but also in the basal plane, can be answered by the following dissociation proposal [*34*]

$$\frac{1}{3}[11\bar{2}0] \rightarrow \frac{1}{9}[11\bar{2}0] + \frac{2}{9}[11\bar{2}0] \quad (5)$$

This leads to a stacking fault ribbon in which the $\{10\bar{1}0\}$ plane of the hcp lattice approximates the $\{112\}$ plane of the bcc lattice. This structural similarity is further improved by the following dissociation, according to Ref *30*.

$$\frac{1}{3}[11\bar{2}0] \rightarrow \frac{1}{6}[12\bar{3}0] + \frac{1}{6}[10\bar{1}0] \quad (6)$$

On the basis of this transformation concept [*34*] and according to a comparative evaluation [*35*], the formation of a stacking fault on the basal or prism planes is linked with the tendency of the hcp lattice to transform into the bcc or fcc structure. Accordingly, an hcp metal slips primarily in the basal or prism planes, depending on whether the ratio of the differences in the free energy is larger or smaller than 1 for the transformation hcp-bcc as opposed to hcp-fcc. This model of the relative stability of the hcp lattice transformation into the bcc or fcc structure [*35*] leads to a consistent prediction of the slip system for all hcp metals with an *a* Burgers vector on a prism or a basal plane, including the temperature and alloying dependences [*30*].

Stacking fault energies for dissociations of *a* dislocations in the prism plane have been measured for titanium, zirconium, and beryllium or estimated via the transformation concept and are likewise tabulated in Table 2.

(*c*) Dissociation of *a* Dislocations in First-Order Pyramidal Planes of Various Modes $\{10\bar{1}n\}$, (n = 1, 2, 3 . . .)

Independent of their *c*/*a* axial ratio, pyramidal planes of the first order and mode $\{10\bar{1}1\}$ represent the most closely packed pyramidal planes. In this planar family, the *a* direction is also included apart from the ($c + a$) direction. A dissociation of the *a* dislocation into partial dislocations is, however, improbable for both these pyramidal planes and for higher order planes owing to the "nearest-neighbor-disturbances."

(*d*) Dissociation of ($c + a$) Dislocations in First-Order Prism Planes and Pyramidal Planes of Various Orders and Modes

In addition to the first-order prism planes $\{10\bar{1}0\}$, the pyramidal planes $\{10\bar{1}1\}$, $\{11\bar{2}1\}$, and $\{11\bar{2}2\}$ also contain the ($c + a$) Burgers vectors [*36*] (Fig. 6).

Whereas a dissociation of the ($c + a$) dislocations in the prism plane is improbable due to the "nearest-neighbor-disturbances," this dissociation is possible in pyramidal planes with energy gain.

Investigations of beryllium [*37*] with slip traces in the $\{11\bar{2}2\}$ plane led to the proposal of a simple dissociation as given by

$$\frac{1}{3}[11\bar{2}3] \rightarrow [0001] + \frac{1}{3}[11\bar{2}0] \tag{7}$$

According to Ref *3*, a dissociation on the $\{11\bar{2}2\}$ plane is described by the reaction (see Fig. 4)

$$(\mathrm{AB} + \mathrm{ST}) \rightarrow \mathrm{SB} + \mathrm{AT} \tag{8}$$

$$\frac{1}{3}[\bar{1}\bar{1}23] \rightarrow \frac{1}{6}[\bar{2}023] + \frac{1}{6}[0\bar{2}23] \tag{9}$$

whereby stacking faults on the basal plane are caused by diffusion (climb).

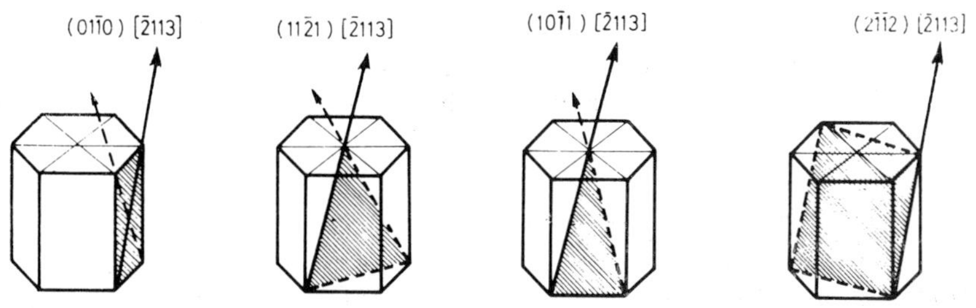

FIG. 6—*Possible slip planes for* (c + a) *Burgers vectors in hcp metals* (*according to Ref* 29).

The atom movements required for $(c + a)$ slip in hcp metals have been discussed by Rosenbaum [*31*] using a hard sphere model for the $\{11\bar{2}2\}$ pyramidal plane. Based on Kronberg's concept [*38,39*] for zonal split, Rosenbaum describes a slip mode on the basis of the formation of a glissile, twin-like stacking fault due to a dissociation into three partial dislocations in three adjacent $\{11\bar{2}2\}$ planes. This model is correspondingly applicable to the $\{10\bar{1}1\}$ and $\{11\bar{2}1\}$ planes. These two plane types can be distinguished from the $\{11\bar{2}2\}$ pyramidal planes by the fact that each plane offers two slip directions with $(c + a)$ type Burgers vector, while the $\{11\bar{2}2\}$ planes contain only one $(c + a)$ type Burgers vector (see Fig. 6). This means that on both of the first-mentioned plane types, twelve slip systems can be operative, whereas on the $\{11\bar{2}2\}$ pyramidal planes, only six ones are possible. Owing to their higher availability of slip systems, their higher atomic density, and their less pronounced waviness (see Fig. 2), the $\{10\bar{1}1\}$ and $\{11\bar{2}1\}$ planes are thus favored energetically compared with the $\{11\bar{2}2\}$ planes.

Thus, the pyramidal planes containing the $(c + a)$ type Burgers vectors offer a means of explaining deformations with c components by slip: it being normally ascribed to twinning. However, the critical resolved shear stresses for $(c + a)$ slip seem to be high, despite the possibility of such deformation configurations occurring.

(*e*) Dissociation of *c* Dislocations on First and Second-Order Prism Planes

The c dislocations are to be found in $\{10\bar{1}0\}$ first and $\{11\bar{2}0\}$ second-order planes. A dissociation on the $(1\bar{2}10)$ plane can be described according to Ref *4* as follows (see Fig. 4)

$$\text{ST} \rightarrow \text{SA} + \text{AT} \tag{10}$$

$$[0001] \rightarrow \frac{1}{6}[\bar{2}023] + \frac{1}{6}[20\bar{2}3] \tag{11}$$

This reaction may generate stacking faults in the basal plane if diffusion (climb) occurs.

In metals with high stacking fault energy, prism and pyramidal slip might be favored at higher temperatures, since the split dislocations coalesce due to thermal activation, thus causing cross-slip [*13*].

2.1.2 Twinning

In hcp metals, twinning (on pyramidal planes of different orders and modes) is usually employed to explain deformations with c components [*1–4,9*].

TABLE 3—*Twinning systems and their elements for various hcp metals according to Ref* 3.

Predicted Twinning Modes				
K_1	K_2	η_1	η_2	in
{10$\bar{1}$2}	{10$\bar{1}\bar{2}$}	⟨10$\bar{1}\bar{1}$⟩	⟨10$\bar{1}$1⟩	Cd, Zn, Mg, Co, Zr, Ti, Be
{22$\bar{4}$1}	{0001}	⟨1, 1, $\bar{2}$, $\overline{12}$⟩	⟨11$\bar{2}$0⟩	...
{10$\bar{1}$1}	{10$\bar{1}$3}	⟨10$\bar{1}\bar{2}$⟩	⟨30$\bar{3}$2⟩	Mg
{10$\bar{1}$1}	*i*	*i*	⟨41$\bar{5}$3⟩	Mg
{20$\bar{2}$1}	{0001}	⟨10$\bar{1}\bar{4}$⟩	⟨10$\bar{1}$0⟩	...
{11$\bar{2}$1}	{0001}	⟨11$\bar{2}$6⟩	⟨11$\bar{2}$0⟩	Zr, Ti, graphite, Re
{10$\bar{1}$3}	{10$\bar{1}$1}	⟨30$\bar{3}$2⟩	⟨10$\bar{1}$2⟩	Mg
{10$\bar{1}$3}	*i*	*i*	⟨2$\bar{1}\bar{1}$3⟩	Mg
i	*i*	*i*	*i*	Mg
{10$\bar{1}$3} Double Twinning				
{13$\bar{4}$0}	{$\bar{1}$100}	⟨7$\bar{5}\bar{2}$0⟩	⟨11$\bar{2}$0⟩	...
{13$\bar{4}$1}	{$\bar{1}$101}	*i*	⟨11$\bar{2}$0⟩	...
{13$\bar{4}$2}	{$\bar{1}$102}	*i*	⟨11$\bar{2}$0⟩	...
{22$\bar{4}$3}	{0001}	⟨11$\bar{2}$4⟩	⟨11$\bar{2}$0⟩	...
{10$\bar{1}$4}	{10$\bar{1}$0}	⟨20$\bar{2}$1⟩	⟨0001⟩	Mg
{11$\bar{2}$2}	{11$\bar{2}$4}	⟨11$\bar{2}$3⟩	⟨22$\bar{4}$3⟩	Zr, Ti
{11$\bar{2}$4}	{11$\bar{2}$2}	⟨22$\bar{4}$3⟩	⟨11$\bar{2}$3⟩	Mg, Ti
{30$\bar{3}$4}	...	...	...	Mg
{11$\bar{2}$3}	...	...	...	Zr, Ti

Most hcp metals form twins of the {10$\bar{1}$2} type, so that the {10$\bar{1}$2} twinning system is often referred to as "normal" twinning. Furthermore, in those metals with a below-normal axial ratio, so-called "abnormal" twin systems occur that exploit first and second-order pyramidal planes of different modes as twinning planes. A list of the twinning modes observed in various metals is given in Table 3. The elements of some systems are completely determined, whereas others are only partly determined or merely predicted.

As seen in Table 3, various twinning modes can become operative, depending on the metal. This is, as with slip processes (see Section 2.1.1), mainly due to:

1. various c/a axial ratios,
2. packing densities and interplanar spacings thereby influenced, and
3. various stacking fault energies and correlated transformation behavior of the hexagonal structure.

The elements of twinning may be described geometrically by means of the reference sphere (Fig. 7). During twinning, the northern half of the reference sphere is deformed into a partial ellipsoid of the same volume. This is accompanied by a homogeneous shear of the crystal lattice parallel to the equatorial plane in such a way that the atom layers form mirror images of each other with respect to this plane. The shear associated with twinning leaves two lattice planes undistorted, that is, all distances and angles in these planes remain unchanged. One such plane is K_1 (the equatorial plane in Fig. 7), which does not change its position with shear. The other plane is K_2 before twinning, K_2' after twinning. All crystal directions in the upper-left segment between K_1 and K_2 are shortened by the shear. All crystal directions to the right of this are extended. The first and second undistorted planes, K_1 and K_2, are normal to the shear plane. The line of intersection of the shear plane with K_1 has the direction η_1, with K_2 the direction η_2, and K_2', the direction η_2'. The shear direction, η_1, is thus parallel to the twinning plane and the magnitude of the homogeneous shear is

$$S = \overline{AA'}, \overline{BB'} \tag{12}$$

where K_2' is related to the magnitude of shear, s, by the angle 2Φ, which K_2, respectively, K_2' makes with K_1 according to

$$S = 2 \cot 2\Phi \tag{13}$$

Thus, the shear magnitude is determined by the crystallographic relationship between the two undistorted planes. Therefore, the twinning mode can be determined completely by K_1 and η_2 or by K_2 and η_1.

The following paragraphs show that there are important differences between the deformation by slip and by twinning.

(*a*) Twinning Plane

Twinning occurs on planes that are not necessarily slip planes.

(*b*) Orientation Changes

Due to the spontaneous transformation, the crystallographic orientation is changed abruptly by a certain angle, ζ. The rotation pole is normal to the shear plane, where the rotation angle is given by

$$\zeta = 2\,(90 - 2\Phi) \tag{14}$$

The change in orientation depends on the respective twinning mode and the c/a axial ratio and may achieve values beyond 90° (values for zirconium; see Section 2.2.2 and Table 4). By contrast during slipping, the orientation above and below the slip plane remains unchanged.

(*c*) Homogeneous Shear

During twinning, the atomic planes parallel to the twinning plane shift over one another by an amount that is a proper fraction of the interplanar spacing in the shear direction. In single lattice structures such as bcc and

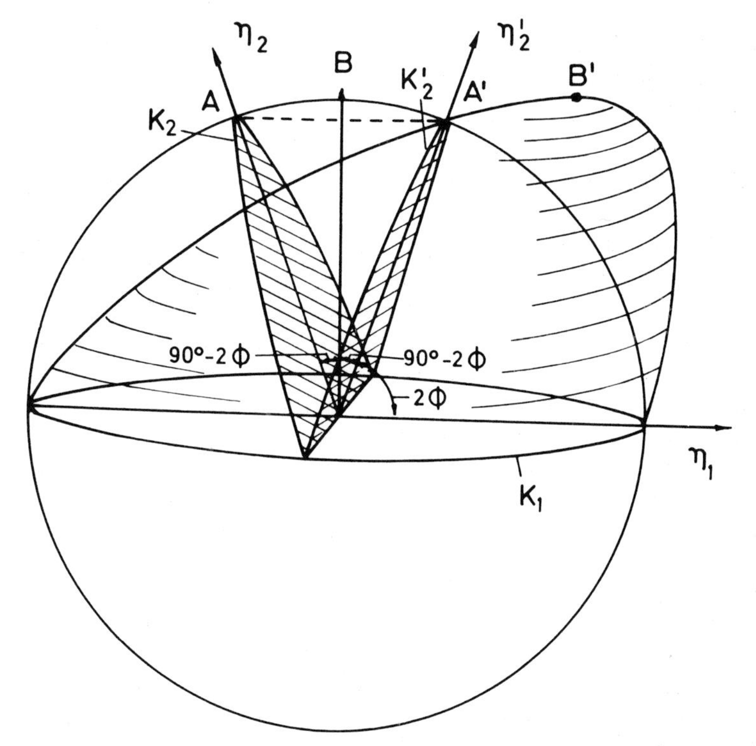

FIG. 7—*Relationship between reference sphere and twinning ellipsoid (according to Ref* 1).

fcc, matrix and twinning are mutually linked by homogeneous shear. In double lattice structures, such as hcp metals, additional minor atomic movements are necessary along directions other than the shear direction. These movements are called shuffles. They are caused by the varying interplanar spacings and the waviness of the non-basal plane. These movements, although necessary on a microscopic scale, do not change the homogeneous shear at a macroscopic level.

By contrast during slipping, deformation is restricted to discrete slip planes or to separated slip zones. The slip magnitudes are integer multiples of the interplanar spacing in the slip direction.

(*d*) Magnitude of Deformation

The magnitude of deformation, ϵ, accompanying twinning is dependent on the orientation, μ, and on the typical c/a axial ratio of the respective crystal; that is, on the activated twinning mode, on the magnitude of the connected shear, s, and on the volumetric proportion, V, of the transformed lattice. For a given twinning system, the relationship may be described by [*10,40*]

$$\epsilon = V \cdot s \cdot \mu \tag{15}$$

where

ϵ = compressive or tensile strain,
V = volumetric proportion of the twin,
s = shear magnitude of the twin, and
μ = Schmid factor (Section 2.3.1)

While the orientation change due to twinning may be large (Section 2.1.2*b*), the contribution of twinning to the overall deformation is small compared with that of slip. For example, the complete transformation (V = 100%) of a magnesium single crystal into its $\{10\bar{1}2\}$ twin position (s = 0.131) for the most favorable orientation (μ = 0.5) is connected with a strain of only about 6.5% (values for zirconium: see Section 2.2.2 and Table 5). However, twinning is essential to the ductility, since the lattice rotations (probably due to secondary or further twinning) reestablish orientations that are favorable for slip and thus allow high deformation magnitudes (Chapters 3 and 4).

(*e*) Dependence of Deformation on Direction

Whereas slip is geometrically reversible under tension or compression, the twinning modes are dependent on the direction, owing to the crystallographic relationship between the extended or compressed orientations. Whether a certain twinning mode leads to an elongation or a shortening in the *c* direction also depends (apart from the crystallographic orientation relationship) on the c/a axial ratio. This is represented by the example of $\{10\bar{1}2\}$ twinning in Fig. 8. Twinning under compression parallel to the *c* axis (shortening) is favored for $c/a > \sqrt{3}$ and for $c/a < \sqrt{3}$ under stress parallel to the *c* axis (elongation) [*4*]. For $c/a = \sqrt{3}$, $\{10\bar{1}2\}$, twinning is impossible [*41*]. Similar considerations may be applied to other twinning modes.

(*f*) Dissociation of Dislocations

For both the explanation of twinning and slip, a dislocation mechanism is employed. However, the dislocations are not perfect, but are so-called twin dislocations that are related to partial ones. Possible deformation models for twinning in hexagonal metals have been discussed by Westlake [*42*] and Rosenbaum [*31*] as well as by Chyung and Wei [*43*].

(*g*) Critical Resolved Shear Stress

The critical resolved shear stress for twinning is not only dependent on the line tension of the initial dislocation, as in the case of slip, but is also dependent on the surface tension of the twin interfaces. It may therefore be assumed that the critical resolved shear stress for twinning—should it even exist [*3*]—is higher than that for slip (see also Section 2.3.2).

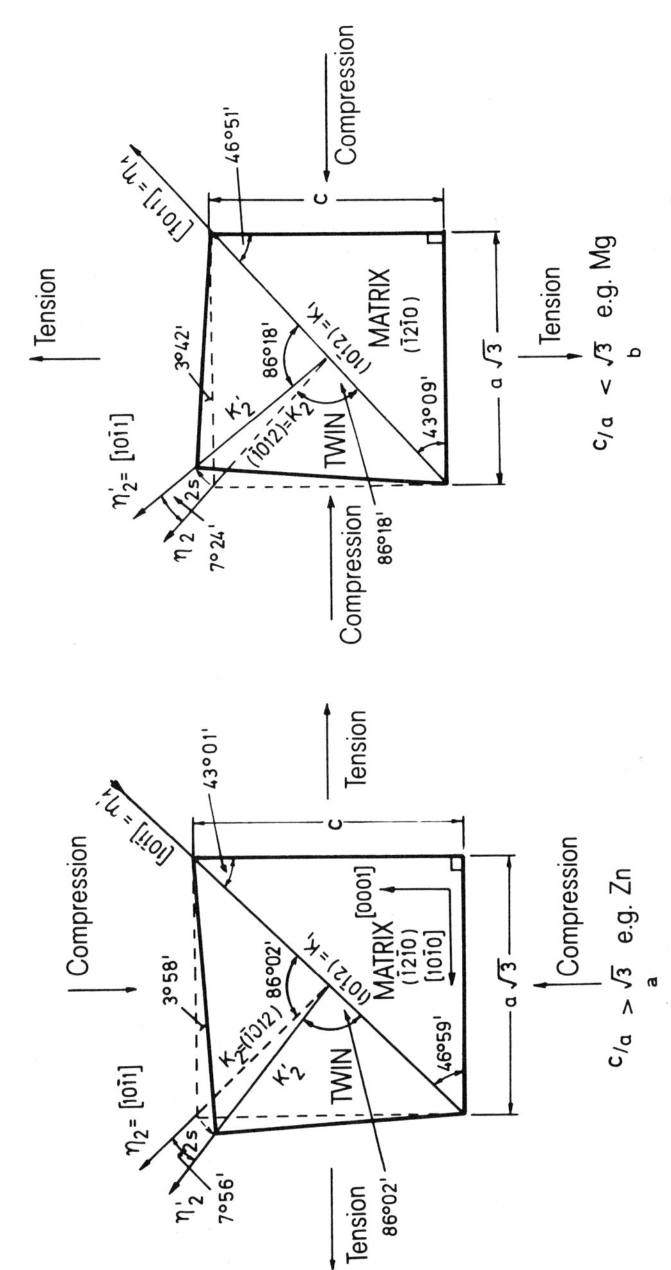

FIG. 8—*Dependence of deformation direction on* c/a *axial ratio in the example of* $\{10\bar{1}2\}$ *twinning for* $c/a > \sqrt{3}$ *(Fig. 8a) and for* $c/a < \sqrt{3}$ *(Fig. 8b) according to Ref 4).*

2.1.3 Kink Bands

The formation of kink bands is another deformation mechanism that occurs especially in hexagonal metals [*44,45*] such as magnesium [*46*], zinc [*12*], cadmium [*47*], titanium [*48–50*], and zirconium [*40,51,52*]. Furthermore, similar kink bands are also formed in fcc metals such as aluminum, molybdenum, and α-uranium [*53–56*].

The kink bands in cadmium rods observed by Orowan [*47*] under compressive deformation along the direction of the rod axis are illustrated in Fig. 9, together with a corresponding schematic sketch.

With the formation of a kink band—as in the case of twinning—a part of the crystal lattice is transferred into another crystal orientation. This process is not thermally activated. In contrast to twinning, however, the kink bands develop gradually rather than in bursts under increasing deformation, whereby neither the new orientation nor the plane of the kink band, *kk'*, have a particular relationship with the initial lattice. The kink plane can be a $\{11\bar{2}0\}$ tilt plane [*52*] that contains dislocations with *a* type Burgers vectors. However, kink planes can be found to contain $\{10\bar{1}0\}$ planes and are often interrelated with twins [*57–59*]. These accommodation kink bands contain two different *a* type Burgers vectors.

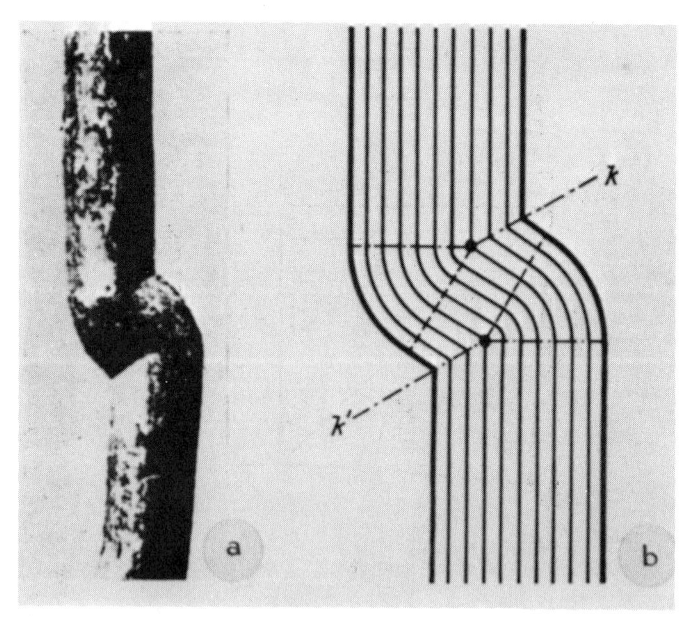

FIG. 9—*Kink bands in a cadmium single crystal* (*according to Ref* 40).

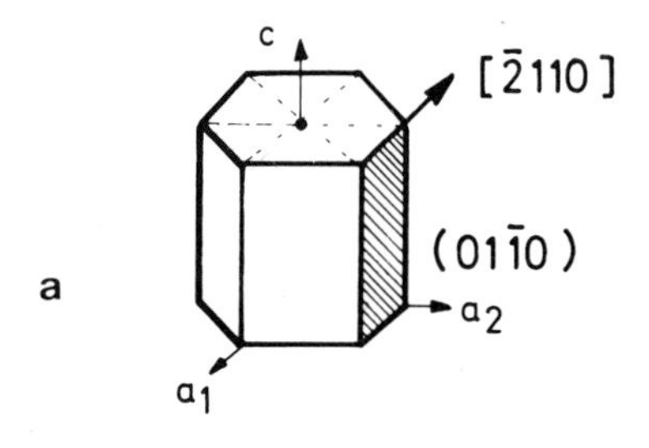

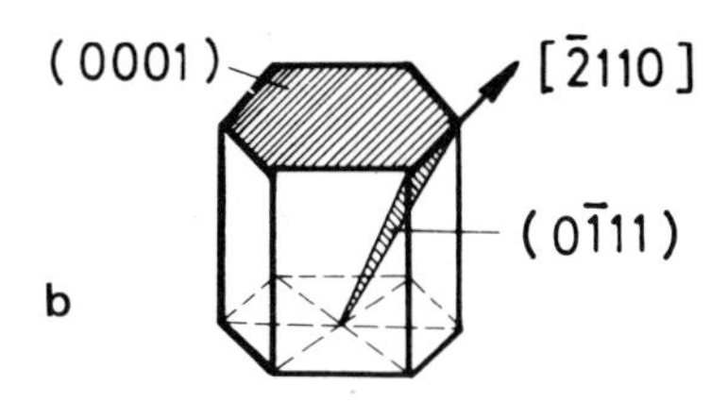

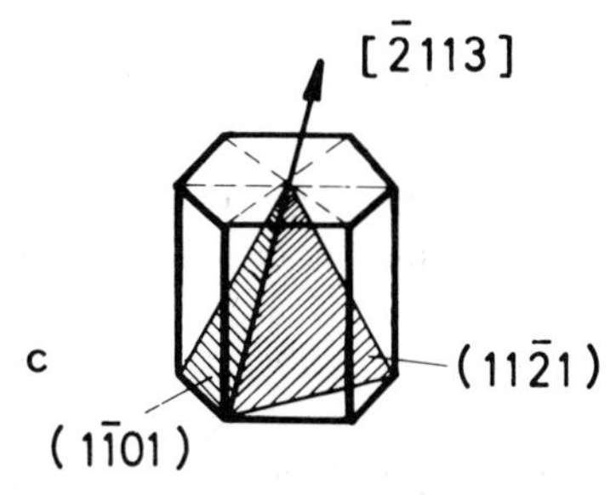

FIG. 10—*Slip systems in α-zirconium (according to Ref* 36).

2.2 Deformation Mechanisms in Zirconium and Zircaloy

2.2.1 Slip Modes

In the α structure of zirconium slip takes place (at least at room temperature and higher, up to about 500°C) usually on the {10$\bar{1}$0} first-order prism planes along the ⟨1$\bar{2}$10⟩ *a* direction [*14,40,60–67*] (Fig. 9*a*). Slip has also been observed in the same direction on the (0001) basal plane [*40,67,68*] (Fig. 9*b*). In regions of high stress concentration, such as grain boundaries, {10$\bar{1}$1} slip traces occur [*62,66,69*] (Fig. 10*b*). A slip system with a component in the *c* direction has been observed only under constraint and at high deformation temperatures. The corresponding slip mode is slip that occurs on first and second-order pyramidal planes {10$\bar{1}$1} and {11$\bar{2}$1} in a $(c + a)$ direction [36,70,71] (Fig. 9c). Further slip planes with less well-defined slip directions can also be given: $\{10\bar{1}2\}$, $\{10\bar{1}3\}$ [64] and $\{11\bar{2}2\}$ [62,66].

For the following reasons, slip on non-basal planes with at least one c component has been asked for several times to explain the observed ductility in zirconium and other hcp metals.

The primary slip system operating in zirconium is prism slip on $\{10\bar{1}0\}$ planes along $\langle 1\bar{2}10 \rangle$ directions. It thus contains no elements that allow deformations in a c direction. Moreover, prism slip contains only two mutually independent deformation modes. However, because zirconium exhibits a high ductility, other deformation mechanisms must be present that fulfill the von Mises compatibility criterion (Section 2.3.5). Since grain boundary slip and twinning are considered to be insufficient [72,73], due to their small contributions to the deformation, prism slip with $(c + a)$ type Burgers vectors has been repeatedly employed to explain the observed ductility [10,74,75,76].

Furthermore, it is well-known that at high temperatures, thermally activated slip processes are favored at the expense of twinning (a non-thermal transformation) whereas the ductility of zirconium increases with increasing deformation temperature [2–4,12,13].

Moreover, the texture of rolled zirconium and its stability at high deformation rates cannot be explained solely by prism slip. Certainly, a complicated sequence of different newly activated twin sequences could account for the observed texture, although an easier alternative explanation would exploit $(c + a)$ type slip [9].

These considerations would allow the existence of $(c + a)$ type Burgers vectors to be presupposed, which could then be verified afterwards by special single-crystal shear experiments [36]. Since it was assumed that $(c + a)$ type Burgers vectors would be operative only under relatively high critical resolved shear stresses, zirconium single crystals were constrained to deform in a given pyramidal plane in the $(c + a)$ direction. These experiments were carried out at different temperatures in order to include the additional effect of the thermal activation. Slip planes and Burgers vector were determined by transmission electron microscopy of thin metal foils taken from the shear region of the single crystal [36].

The slip plane can be determined by the trace of the dislocation band on the contrast image (Fig. 11a) and by the transformation into the stereographic projection of the corresponding foil plane (Fig. 11b) without distorting the angular relationship. The foil plane and zonal axis of the slip plane can be derived from the electron diffraction pattern by taking into account the foil thickness.

The Burgers vectors, forming the single dislocations of the dislocation band, can be determined by contrast experiments performed under two-beam conditions with displacement of the diffracting set of net planes (Fig.

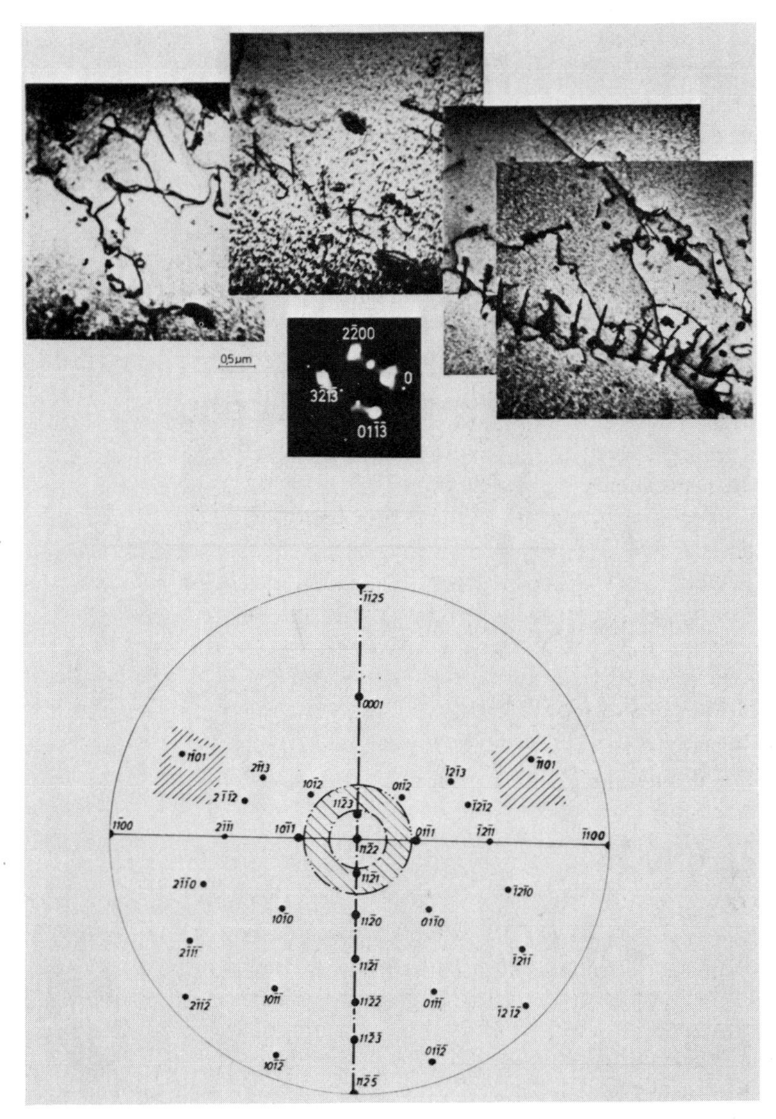

FIG. 11—*Dislocation band in contrast image* (a) *and transformation into the stereographic projection* (b) *of the corresponding foil plane result in the determination of the slip plane* (*according to Ref* 36).

12*a*). As dislocations lying in the reflecting net plane do not afford any contrast, the corresponding Burgers vector can be found according to the $g \times b = 0$ criterion [*77,78*], (Fig. 12*b*), whereby *g* is the lattice vector of the diffracting set of net planes and *b* is the Burgers vector under consideration.

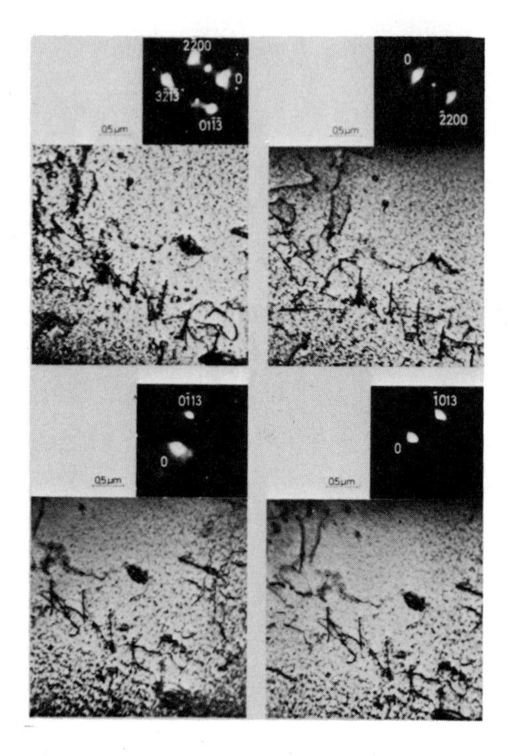

REFLECTION g	BURGERS VECTORS b OF THE PERFECT DISLOCATIONS (x 1/3)									
	a TYPE			$(a + c)$ TYPE						c TYPE
	$\pm[11\bar{2}0]$	$\pm[\bar{1}2\bar{1}0]$	$\pm[\bar{2}110]$	$\pm[11\bar{2}3]$	$\pm[\bar{1}2\bar{1}3]$	$\pm[\bar{2}113]$	$\pm[11\bar{2}\bar{3}]$	$\pm[\bar{1}2\bar{1}\bar{3}]$	$\pm[\bar{2}11\bar{3}]$	$\pm[0003]$
$\bar{2}200$	0	±2	±2	0	±2	±2	0	±2	±2	0
$\bar{1}013$	∓1	0	±1	±2	±3	±4	∓4	±3	∓2	±3
$0\bar{1}13$	∓1	∓1	0	±2	±2	±3	∓4	±2	∓3	±3
$\bar{2}112$	∓1	±1	±2	±1	±3	±4	∓3	±3	0	±2

FIG. 12—*Contrast images under two-beam conditions* (a) *and the corresponding values of* g · b *for the various perfect dislocations* (b) (*according to Ref* 36).

The results of these investigations show that a $(c + a)$ type Burgers vector can be operative in zirconium, that is, from room temperature to 75°C, and its occurrence is more likely the higher the temperature. The $\{11\bar{2}1\}$ or $\{10\bar{1}1\}$ pyramidal planes proved to be slip planes for $(c + a)$ dislocations, that is, those planes that seem to be favored energetically from a theoretical viewpoint (see Figs. 2 and 6 and Section 2.1.1.2).

Apart from the pyramidal slip modes, deformation with *c* components have been explained primarily by twinning on first and second-order pyramidal planes.

2.2.2 Twinning Modes

Under tensile stress along the direction of the *c* axis, $\{10\bar{1}2\}\langle\bar{1}011\rangle$ twins are activated [*28,40,61–63,65,69,79–81*] and less commonly $\{11\bar{2}1\}\langle\bar{1}\bar{1}26\rangle$ twins [*28,40,60–63,65,66,69,80–82*] (Fig. 12*a*). Under compression in a *c* direction $\{11\bar{2}2\}\langle\bar{1}\bar{1}23\rangle$ twinning [*40,60,62,63,65,66,69,80,81*] and at elevated temperatures $\{1011\}\langle1012\rangle$ twinning [*70,81*] are observed (Fig. 13*b*). For some cases, an as yet incompletely defined twinning mode on the $\{11\bar{2}3\}$ compound plane is mentioned [*40,60–63*].

The determination and definition of the four elements forming the respective twinning mode are illustrated by the example of the $\{10\bar{1}1\}$ twinning system (see Fig. 14), which becomes operative in zirconium at elevated temperatures [*81*]. The respective orientations, which can be taken from

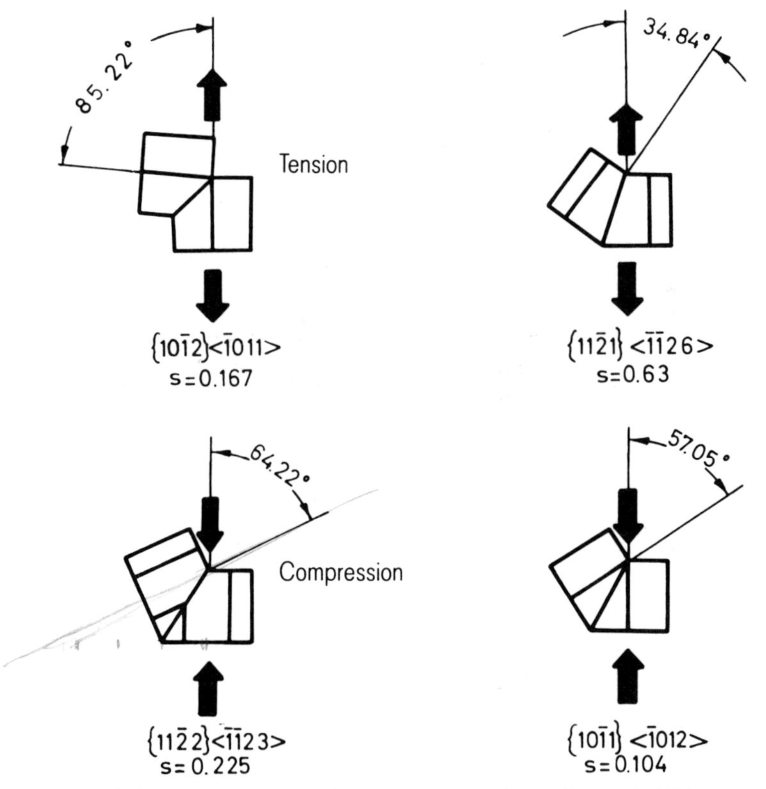

FIG. 13—*Twinning modes in α-zirconium (according to Ref* 81).

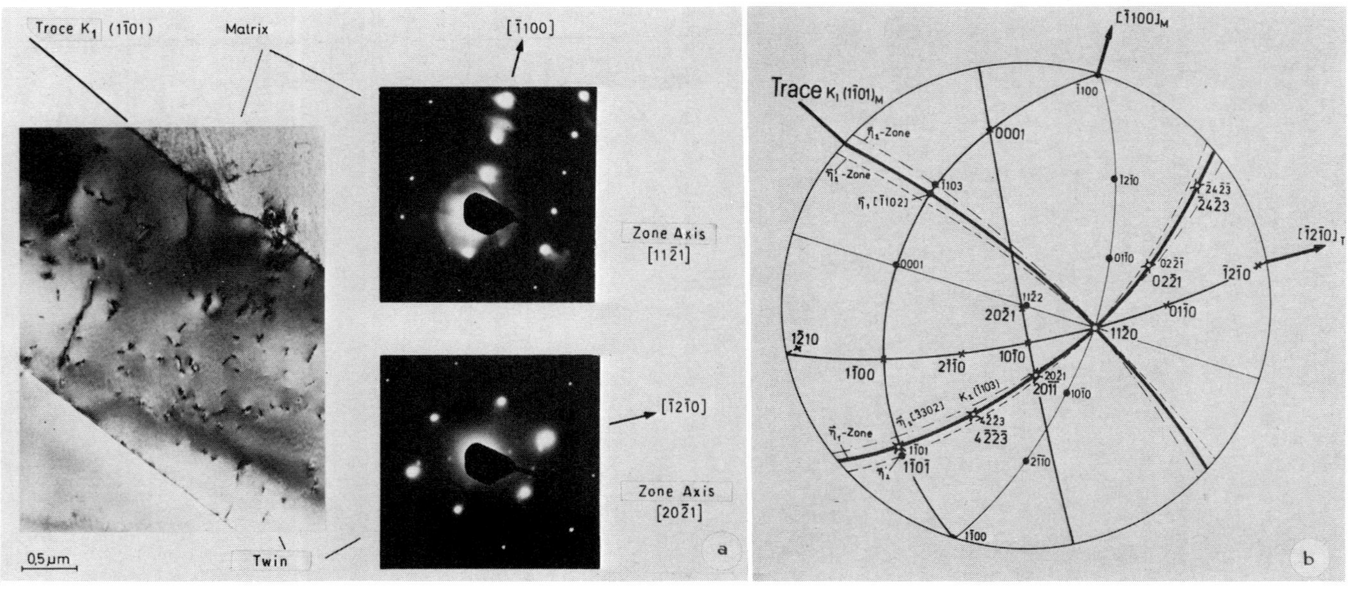

FIG. 14—*Contrast image, electron diffraction pattern (Fig. 14*a*), and the derived stereographic projection (Fig. 14*b*) of the orientation relationships between matrix and ($1\bar{1}01$) twin:* ● = *normal to the plane in the matrix;* + = *normal to the plane in the ($1\bar{1}01$) twin,* –⌖– = *normal to the plane in coincidence for matrix and twin; thin lines = zones of the matrix; thick lines = zones of the twin; all indices refer to the matrix (according to Ref* 81).

the contrast image and the corresponding diffraction lattices of matrix and twin, can be transformed into the stereographic projection (Fig. 14*b*). Both orientations obey the laws mentioned in Section 2.1.2, which in this case apply to a compound system of a double, rotationally symmetric lattice. The poles of the twin lattices can be brought to coincide with the matrix lattice by a rotation of 57.05° about the pole of the shear plane, which is the $(11\bar{2}0)$ plane in this example. The shear direction, $\eta_1 = [\bar{1}102]$, is fixed by the point of intersection between the twin trace, K_1, and the trace of the shear plane.

The following criteria:

1. smallest possible magnitude of shear due to net displacement of atoms,
2. rational indices due to high lattice symmetry, and
3. coincidence of the lattice due to reflection in K_1 as well as by rotation about η_1,

result in the further twin elements K_2 $(\bar{1}103)$ and η_2 $[\bar{3}302]$. From this, the magnitude of shear, $s = 0.1044$, can be derived. The pole of the shear plane and the pole of the twin plane lie on the zone that contains the poles of equal indices for matrix and twin (Fig. 14).

The twinning systems observed in zirconium, the corresponding twin elements as well as the lattice rotation and the magnitude of shear are summarized in Table 4, according to Eqs 13 and 14. Table 5 gives a list of the maximum magnitudes of deformation due to twinning for the twinning modes operative in zirconium according to Eq 15.

Thus, also for zirconium, twinning does not allow high rates of deformation. However, it may play an important role in deformation, since lattice regions initially unfavorable for slip are reoriented more favorably by the lattice rotation, thus allowing the possibility of larger strains. Moreover, it should be noted that twinning activates new slip systems even in regions of local stress concentration such as twin boundaries or by interactions between dislocation and twinning. This possibility has been discussed especially for slip systems with $(c + a)$ Burgers vector [*2,83–86*].

2.3 Factors Influencing the Deformation Mechanisms

In order to predict if slip or twinning occurs or, more precisely, which of the deformation modes will be activated, the following factors must be taken into account.

2.3.1 Schmid Factor

The Schmid factor [*16*] is a purely geometrical relationship between the corresponding deformation mode and the direction of force. The relation-

TABLE 4—*Twinning systems and their elements in zirconium* $\{10\bar{1}2\}$, $\{11\bar{2}1\}$, $\{11\bar{2}2\}$, *and* $\{10\bar{1}1\}$ *for* c/a *axial ratio* = *1.5931 (according to Ref* 81).

Twinning Plane, K_1	Shear Direction of Twinning, η_2	Second Undistorted Plane, K_2	Direction of Line of Intersection Between K_2 and the Shear Plane, η_2	Shear Plane Normal to K_1 and K_2	Magnitude of Shear, S	Lattice Rotation, ξ	Ref
$\{\bar{1}012\}$	$\langle\bar{1}011\rangle$	$\{\bar{1}012\}$	$\langle10\bar{1}1\rangle$	$\{1\bar{2}10\}$	0.167	94.87°	*61* to *63*
$\{11\bar{2}1\}$	$\langle\bar{1}\bar{1}26\rangle$	$\{0001\}$	$\langle11\bar{2}0\rangle$	$\{\bar{1}110\}$	0.63	34.84°	*61* to *63*
$\{11\bar{2}2\}$	$\langle\bar{1}\bar{1}23\rangle$	$\{\bar{1}\bar{1}24\}$	$\langle22\bar{4}3\rangle$	$\{1\bar{1}00\}$	0.225	64.22°	*61* to *63*
$\{10\bar{1}1\}$	$\langle\bar{1}012\rangle$	$\{\bar{1}013\}$	$\langle30\bar{3}2\rangle$	$\{1\bar{2}10\}$	0.1044	57.05°	*70*

TABLE 5—*Attainable strains ϵ in % (maximal or parallel to* c *axis) for zirconium twinning systems, assuming a complete transformation into twin position;* (+) = *tension and* (−) = *compression* ∥ c.

Twinning Mode	Magnitude of Shear	ϵ, % for μ_{max} = 0.5	Loading ∥ *c* axis	
			μ	ϵ, %
$\{10\bar{1}2\}$	0.167	+8.35	0.497	+8.3
$\{11\bar{2}1\}$	0.63	+31.5	0.278	+17.5
$\{11\bar{2}2\}$	0.225	−11.25	0.45	−10.15
$\{10\bar{1}1\}$	0.1044	−5.22	0.442	−4.62

ship between the resolved shear stress operative on the slip plane in the slip direction, the external direction of force, and the specimen dimension is given in Fig. 15

$$\tau = \frac{F}{A} \cdot \cos\phi \cos\lambda \tag{16}$$

μ is the so-termed Schmid factor

$$\mu = \cos\phi \cos\lambda \tag{17}$$

where

τ = resolved shear stress (rss) on the slip plane in the slip direction,
F = externally applied force in the direction of the rod axis,
A = cross section of the crystal,
ϕ = angle between the normal to the slip plane and the direction of force of the external reference system, and
λ = angle between the slip direction and net force of the external reference system.

This relationship applies correspondingly to twinning modes with:

τ = resolved shear stress on the twinning plane in the shear direction,
ϕ = angle between the normal to the twinning plane, K_1, and the direction of force of the external reference system, and
λ = angle between the shear direction of twinning, η_1, and the direction of force of the external reference system.

In hcp metals, the resolved shear stresses of the various deformation modes are strongly dependent on the direction of force, especially with respect to the *c* axis. The orientation dependence is illustrated by Fig. 16, which shows the Schmid factors of important slip and twinning modes operative in zirconium for the range of angles from 0 to 90° between the basal pole and the direction of force.

Owing to the low offer of slip systems, the hcp metals with certain orientations show a pronounced tendency to geometrical hardening or softening. This behavior is caused solely by a change in the Schmid factor, due to a lattice rotation during the slip process, and is independent of hardening by dislocation reactions [*15*].

2.3.2 Critical Resolved Shear Stress, $\tau_{(crss)}$

For a given deformation system, slip takes place if the acting resolved shear stress, $\tau_{(rss)}$, exceeds a critical value. This value is termed critical resolved shear stress, $\tau_{(crss)}$, and constitutes a material property that may be influenced by temperature, impurities, as well as by the mechanical and thermal history (rate of deformation and of recrystallization).

In various hcp metals [*63,87,88*], single-crystal experiments were performed in order to obtain critical resolved shear stresses for basal and

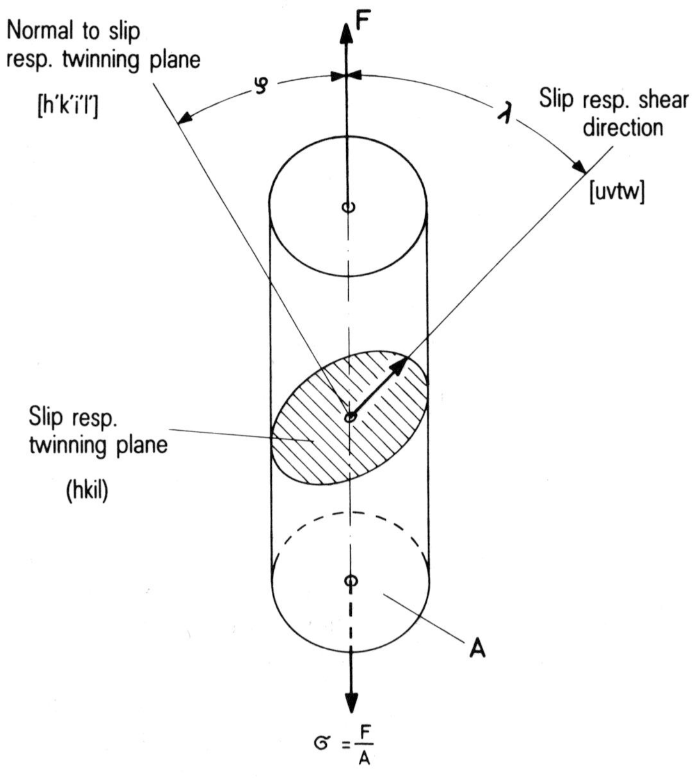

FIG. 15—*Relationship between the shearing stress acting in the deformation system and the force applied externally on the single-crystal rod. ([h′ k′ i′ l′], because in hcp lattices the planes and directions of the same Miller-Bravais indices are not necessarily normal to each other. This is valid for all pyramidal planes; exceptions are basal and prism planes).*

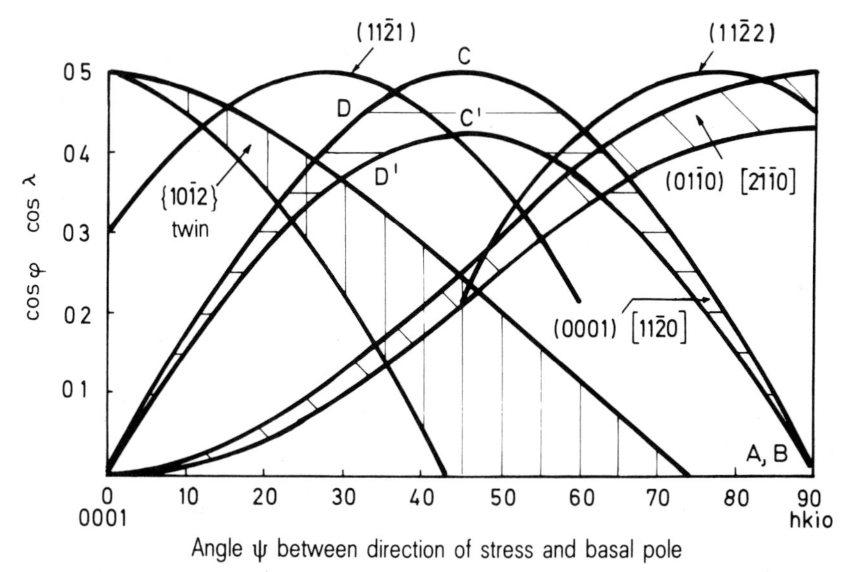

FIG. 16—*Schmid factor orientation dependence for various slip and twinning systems in zirconium* (*according to Ref* 40).

prism slip on a macroscopic scale. In addition to the aforementioned parameters, these values are strongly dependent on the testing method and the accuracy of the measurement, so that the critical values cited in the literature differ widely.

For zirconium, critical resolved shear stresses have been determined exclusively for prism slip systems with *a* type Burgers vectors [*60–63*]. In Table 6, the values for various deformation temperatures are listed. Data on critical resolved shear stresses of pyramidal slip systems are given neither for zirconium nor for other hcp metals.

Values of corresponding critical stresses that initiate the various twinning modes cannot as yet be given, because the mechanisms governing the

TABLE 6—*Critical resolved shear stresses,* $\tau_{(crss)}$, *for prism slip in zirconium at various deformation temperatures where* (c) = *measured under compression and* (t) = *measured under stress.*

Deformation Temperature, °C	$\tau_{(crss)}$, N/mm²	Ref
−197	9.8 (t)	*63*
RT	6.4 (c)	*60, 61, 62*
302	2 (t)	*63*
802	0.2 (c)	*63*

TABLE 7—*Relative ratios of the derived critical shear stresses for various deformation systems in zirconium and Zircaloy after deformation at room temperature (according to Ref* 91).

Material	Form	Loading Condition at Room Temperature	$\frac{crss_{\{10\bar{1}2\}\ twin}}{crss_{\{10\bar{1}0\}\ slip}}$	$\frac{crss_{\{11\bar{2}1\}\ twin}}{crss_{\{10\bar{1}0\}\ slip}}$	$\frac{crss_{\{11\bar{2}2\}\ twin}}{crss_{\{10\bar{1}0\}\ slip}}$	Ref
hcp (Zr)	...	...	1.5	...	2	*92*
Zircaloy-4	sheet	uniaxial	1.55	...	2.03	calculated after data of Ref *93*
Zircaloy-4	tube	biaxial	1.25	1.35	1.45	*94*

generation and the growth of twinning, that is, whether slip precedes the generation of twinning, have yet to be resolved [*3,89,90*].

For polycrystalline materials for which the yield strength or the 0.2% offset yield strength is given in place of the critical resolved shear stress, corresponding values can be derived for slip as well as for twinning by, for example, tensile or compressive tests of textured specimens. Table 7 gives some independently determined test results of zirconium and Zircaloy. In order to eliminate the material and testing inaccuracies, the respective ratios of the derived critical shear stresses are listed for various deformation modes [*91*].

Table 7 shows that the critical resolved shear stresses and the corresponding stresses initiating the various twinning modes are different within the same material, but are higher than those for prism slip in every case (see Section 2.1.2*g*). However, the deformation system with the lowest critical shear stress need not necessarily be operative, since the Schmid factor is relevant for the acting resolved shear stress and is, especially in hcp metals, strongly dependent on the position of the basal pole with respect to the direction of force (see Section 2.3.1 and Fig. 10).

2.3.2.1 Dependence on Impurities and Alloying Elements—The critical resolved shear stress required to initiate the corresponding deformation mode may be influenced by impurities and alloying elements in the form of substitutional and interstitial solid solutions as well as in the form of decomposed and precipitated phases. The decisive factors are: changes in interplanar spacing and atomic packing density, changes in stress field due to lattice distortions, and changes in stacking fault energy accompanied by the possible coalescence of dissociated screw components (cross-slip at a high stacking fault energy).

Conrad et al. [*95*] showed by experiments on the deformation kinetics of α-titanium that the strengthening effects of substitutional solid solutions on single crystals and textured polycrystals is caused by the fact that the critical resolved shear stress is increased due to differences in atom sizes and due to changes in bond energies, although the slip system is retained. (Conrad [*96*] assumes that this result confirmed for prism slip also applies to the other deformation modes possible in hcp metals.)

Additionally, it can be seen from Table 7 that all ratios of the stresses initiating the various twinning modes to the critical resolved shear stress for prism slip are more or less maintained, although the absolute values can be increased by, for example, alloying and deformation.

Taking into account these relationships, it seems justified that the knowledge gained about the deformation mechanisms from zirconium may also be applied to zirconium-rich alloys, such as Zircaloy. Therefore it is possible to explain with the same mechanisms the data on deformation textures and mechanical anisotropy obtained from the alloy.

2.3.2.2 Dependence on the Deformation Temperature—The previously discussed results apply to temperatures around room temperature. However, in regions above and below these temperatures, certain deformation modes may not operate or new ones may be activated.

Generally, twinning is preferred to slip at low deformation temperatures; at higher temperatures, however, slip is favored. In this case, the decisive factors may be thermally activated processes, such as the surmounting of stress fields caused by the impurity atoms and the coalescence of screw components for cross-slip. These processes may act differently according to the degree of impurity and deformation.

2.3.2.3 Deformation Direction—For twinning, the dependence of the various systems on the macroscopic change in shape must also be taken into account, that is, the twinning mechanisms are dependent on the direction of deformation (see Sections 2.1.2*e* and 2.2.2).

For a single crystal under uniaxial loading, these items (2.3.1 to 2.3.3) would be sufficient to predict quantitatively the operating deformation system. If—as with all hcp metals—twinning was activated at certain orientations, it should be possible to predict qualitatively the corresponding operating deformation systems on the basis of the aforementioned estimations of the critical stress initiating twinning.

For technically applied materials, that is, polycrystalline materials employed under operating conditions, the predictions are much more complicated, because the following factors must be taken into account.

2.3.4 Multiaxial Stress Conditions

Technically applied materials are normally subjected to multiaxial loadings. Thus, it is possible only in some cases to quantify the loading conditions accurately. For example, for the relatively simple process of sheet rolling, the forces acting during deformation cannot be given precisely [*97,98*]. For deformation processes such as tube drawing and rocking, the interaction of forces is even more complicated.

Apart from the problem of determining multiaxial forces exactly, there is another difficulty even with knowledge of the individual forces due to the anisotropic deformation behavior of metallic materials. The yield strength measured under uniaxial loading can only be applied to predict the yield point under multiaxial loading for homogeneous, isotropic materials. In isotropic plastomechanics, it is possible, for example, knowing the three principal stresses (σ_1, σ_2, σ_3), to compute the onset of plastic deformation, if the yield criterion of von Mises [*99*] is fulfilled by the equation

$$(\sigma_1 - \sigma_2)^2 + (\sigma_2 - \sigma_3)^2 + (\sigma_3 - \sigma_1)^2 = 2K_f^2 \qquad (18)$$

that is given for a cylinder of infinite length with the axes passing through the coordinate origin and at an equal angle to all three axes.

The 0.2% offset yield strength determined in a uniaxial tension test is normally used for the deviation of the design stress intensity, K_f.

In order to consider the anisotropic yield behavior of metallic materials, Hill [*100*] has extended the maximum distortion energy criterion according to von Mises by the introduction of anisotropy parameters, without going into the reasons for the anisotropy itself. Propositions to establish the mechanical anisotropy by crystallographic means, that is, by criteria for operative deformation mechanisms, have been made by Taylor [*101*] with the help of the principle of least-work-performed in deformation as well as by Bishop and Hill [*102*] with the equivalent principle of maximum-work-performed in deformation. A precise prediction of the operative deformation mechanism under triaxial tensile loading has yet to be formulated. Initial steps towards computing the mechanical anisotropy under the simplified conditions of uniaxial or biaxial loadings in textured fcc and bcc metals were successfully carried out by Piehler [*103*], Backofen, Hosford et al. [*8,104,105*], and by Althof, Drefahl, and Wincierz [*106*].

For hcp metals, the theory was initially presented by Chin and Mammel [*107*] as well as by Thornburg and Piehler [*108*] according to the principle of maximum-work-performed in deformation. Here, the factors mentioned in Sections 2.3.1 to 2.3.2 complicate both the prediction of the texture development and the prediction of the mechanical anisotropy. Good agreement between theoretically and experimentally derived yield loci, especially for zirconium and Zircaloy, is reported by Hosford [*109*], Dressler, Matucha, and Wincierz [*94,110*].

2.3.5 *Compatibility Conditions*

Assuming a constant volume, there must be, according to the von Mises criterion [95], at least five mutually independent deformation modes in a single grain embedded in a polycrystal in order to ensure a homogeneous deformation of the grain on the macroscopic scale without cracking along the grain boundary. These compatibility conditions, that is, the accommodation of the deformation at the grain boundaries and at similar microstructural features, result in a superposition of inner stresses and ex-

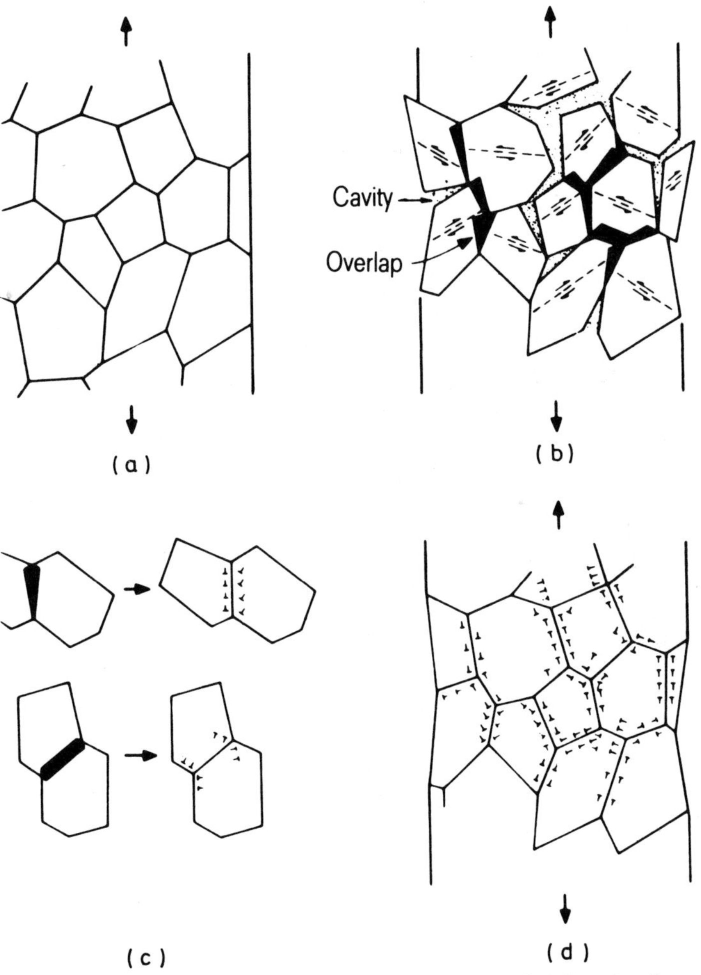

FIG. 17—*Deformation model of a polycrystalline material. If each grain of a polycrystal deforms equally, overlaps and cavities would develop* (17b). *These can, however, be offset by the introduction of dislocations* (17c *and* 17d) (*according to Ashby* [111]).

ternally applied stresses. This effect, which is more pronounced the smaller the grain size and the higher the difference on orientation toward the next grain, makes it difficult to estimate the resulting stresses initiating the deformation system in a single grain. Up to this time, a quantitative determination has been impossible.

For deformation in polycrystalline materials, Ashby [*111*] illustrated the highly complicated interactions of dislocations along grain boundaries (Fig. 17) by arranging dislocations in a model in such a way so that overlaps or cavities are compensated during deformation. The interactions are even more complex if twinning is included in the accommodation process, as with hcp metals.

2.3.6 Preferred Crystallographic Orientation

When predicting the operative deformation mechanisms of polycrystalline materials, the texture must be taken into account, especially for hcp metals that form pronounced deformation and annealing textures due to their variety and distribution of deformation modes. If the crystals were randomly oriented, the anisotropy of each grain would be largely balanced by the random distribution. However, if the grains have a preferred crystallographic orientation, then the properties of the polycrystal approximate those of the single crystal. This behavior is more pronounced the stronger and sharper the texture and the coarser the grain.

As yet, especially for the complicated relationships encountered in hcp metals, there have only been initial attempts at recording these additional factors (Sections 2.3.4 to 2.3.6) quantitatively (Chapter 4).

Chapter 3—Texture Development

3.1 Quantitative Determination of Textures

In order to trace the texture development, as well as to relate the texture, for example, to mechanical anisotropy, an exact and quantitative determination of the preferred crystallographic orientation is required. This requires comparative intensity values because the measured lattice plane intensity of a textured specimen must be related to that of the same lattice plane of a randomly oriented specimen.

Depending on the X-ray techniques used to determine the texture, various factors can cause a loss in intensity, such as defocusing (reflection method) and absorption-plus-defocusing (transmission method), due to movement and geometries of the specimen and the diffractometer. The various techniques and necessary corrections are listed in Ref *6*. For the reflection method proposed by Schulz [*112*], the most widely used technique, several investigations [*113–116*] show that the angle between the normal to the sample and the pole of the diffracting lattice plane results in a strong defocusing effect, which is also dependent on the Bragg diffraction angle. The correction function for the reflection method, which completely describes the geometry, and the physical properties of the measurement equipment, as well as the movement of specimen and goniometer [*116*] is thus given by

$$\frac{I_A\,(\Phi,\,\Theta,\,W_b,\,L_R)}{I_A\,(\Phi\,=\,O,\,\Theta,\,W_B,\,L_R)} = 1 - \frac{2}{(2\pi)^{1/2}} \int_{-\infty}^{-L_R/P(W_B \tan\Phi \sin 2\Theta/\sin\Theta)} e\,\frac{-y^2}{2}\,dy \qquad (19)$$

where

I_A = measured X-ray intensity above background level,
Φ = tilt angle between the normal to the sample surface and the normal to the diffracting net plane,
Θ = Bragg angle derived from $n\lambda = 2d \sin \Theta$,
W_B = width of the incident X-ray, given by the slit width of the aperture,
L_R = slit width of the X-ray counter,
P = a constant factor that is determined empirically for the individual equipment to account for effects such as flatness of the sample, misalignment of the goniometer or the sample, the intensity profile, and the wavelength distribution of the X-ray source (P lies in the order of 1.0).

The integral calculates the intensity loss, y, due to collimation of the broadened, diffracted beam at the counting tube. By means of this convolutionary integral, or rather by a simplified computer numerical version of the equation, a correction of the defocusing effects and hence a quantitative determination of the intensity is possible [*116*]. The latter is required for the classical presentation of textures by pole figures. The drawback to this two-dimensional plot is that many orientations come together at one point in the pole figure. In order to reduce this ambiguity, several pole figures of various crystallographic planes are usually plotted.

In recent years, mathematical methods were developed [*117,118*] to compute three-dimensional orientation distribution functions (ODF) with the help of intensity values derived from several pole figures. This ODF-analysis shows, by averaging, the extent to which crystals of a given orientation are present in a polycrystalline sample and thereby represents a more exact quantification of the texture analysis.

3.2 Theory about the Development of Deformation Textures in HCP Metals

In order to understand the development of deformation textures and to predict ideal preferred orientations, one needs to consider the deformation mechanisms, that is, systems with set crystallographic slip or twinning planes and the corresponding slip or shear directions (see Chapter 2).

The various theories referred to in the literature that explain texture development differ in the combination of five slip systems chosen out of the overall possible combinations according to the yield criterion of von Mises [*99*]. The operative deformation mechanisms chosen by Taylor [*101*] were based on the principle of internal least-work-performed in deformation, while Bishop and Hill [*102*] chose the equivalent principle of external maximum-work-performed in deformation (see Section 2.3.4).

The theories concerning the development of deformation textures have been applied successfully to cubic metals, in particular for deducing deformation textures of fcc metals [*119–121*]. In comparison, the theory regarding the development of deformation textures in hcp metals is not as extensive and complete. There are several reasons for this (see Chapter 2):

(*a*) The slip systems are neither as numerous nor as symmetrically distributed in hexagonal metals as in cubic metals. Therefore, twinning competes with slip and, depending on the conditions of deformation, can play a significant role.

(*b*) The lattice rotations caused by the unipolar twinning modes depend on stringent crystallographic orientation relationships and are thus inhomogeneous. However, the theories of deformation available so far are based on homogeneous deformations.

(*c*) The altogether larger number of deformation systems (slip and twinning) results in significantly more combinations of five mutually independent deformation systems.
(*d*) The interaction between slip and twinning is complicated and makes the formulation of the theory for texture development more difficult.
(*e*) The critical resolved shear stresses (crss), which along with other factors are required by the theories of deformation, are different in hcp metals for the various deformation systems and are not entirely known; for the twinning systems, it is not certain whether they exist at all. As a substitute for these, shear stress ratios, derived from yield strengths of textured polycrystalline materials, have been used.
(*f*) In the hcp lattice, the operative deformation systems vary from metal to metal and are up to now not entirely known. Therefore, the theory of the development of deformation textures must be altered or tailored for virtually every hcp metal.

Approaches to explain the development of deformation textures in hcp metals have been made by Calnan and Clews [*122*] for zinc and magnesium, Williams and Eppelsheimer for titanium [*76*], and by Hobson for zirconium [*123*].

Calnan and Clews [*122*] predicted on a theoretical basis the main rotations of the crystal lattice by averaging the rotation tendencies of the grains for all possible orientations. They assumed that the deformation systems with a maximum shear stress ratio, rss/crss, become effective. Whether slip or twinning occurs also depends, among other factors, on this ratio. In addition, they took into account that there is a difference between applied stress and effective stress due to the deformation accommodations at grain boundaries, etc.

Williams and Eppelsheimer [*76*] relied on this theoretically derived prediction. Taking into account the interactions of those deformation mechanisms known for titanium, they traced the texture development in polycrystalline sheet material by means of (0002) and $\{10\bar{1}0\}$ pole figures.

Of course, these integral methods cannot predict in detail which deformation mechanisms really become effective in the single grains after subsequent deformation steps.

Hobson [*123*] deformed zirconium single crystals of different orientations by rolling and drawing procedures and then traced the lattice rotations caused by the deformation systems. He concluded that twinning is more important in the initial stages and that slip is more important in the final stages of texture development. Under the simplifying assumption that equal critical resolved shear stresses for the different deformation systems, the results are explained by shear stress contour diagrams, but only for the main deformation system with the highest respective Schmid factor. No simultaneous activation of five independent deformation modes has been considered. Furthermore, the experiments started with degrees of defor-

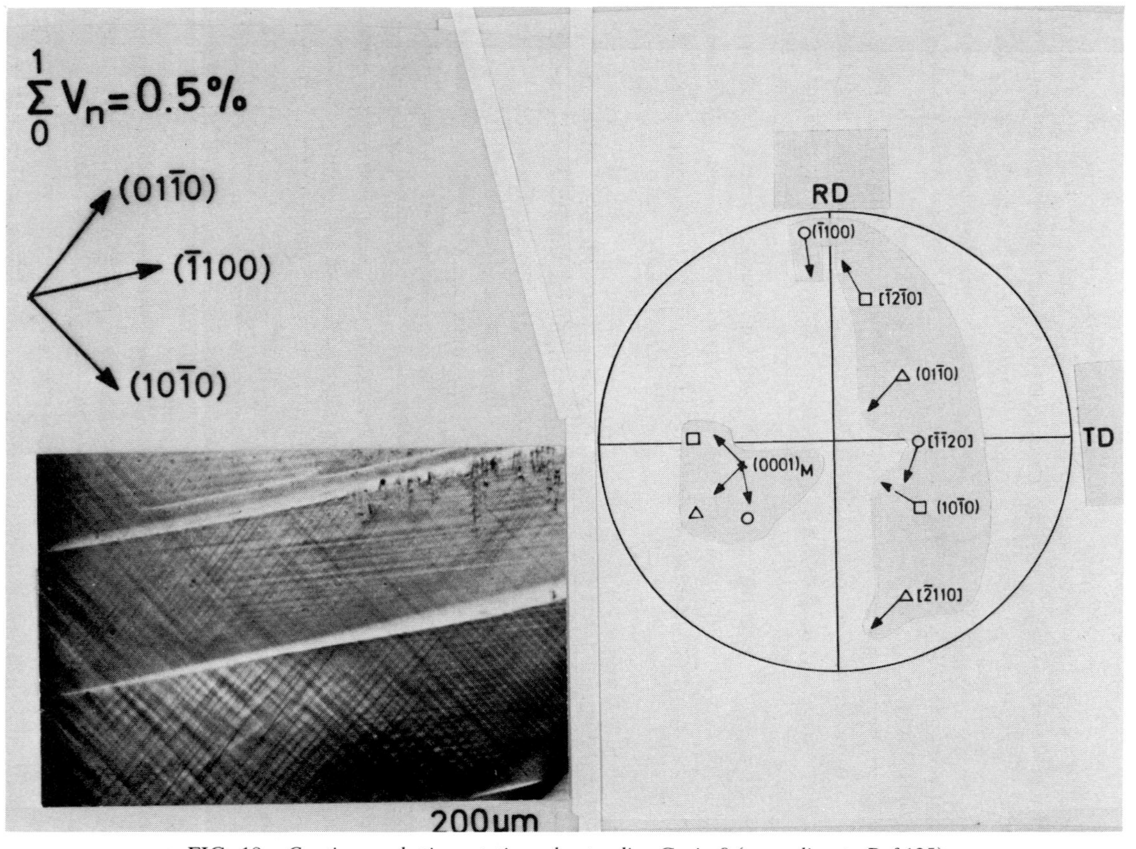

FIG. 18—*Continuous lattice rotations due to slip, Grain 9 (according to Ref* 125).

mation higher than 20%. The texture development below this degree of deformation was not followed in detail. As the results of this study show, this is, however, very manifold and informative.

The main difficulty in applying all these theories is that the critical resolved shear stresses are needed, but they are not or only partly known for the slip and twinning systems in hcp metals.

Therefore, a polycrystalline, coarse-grained zirconium polycrystal with a given, more or less, random distribution of the individual grains was deformed in sequential rolling steps. After each pass the activated slip and twinning systems have been traced throughout each individual grain, correlated to the corresponding lattice rotations, and compared with the integral texture measurement [*124,125*].

The results show that the criterion of compatibility according to von Mises is fulfilled even in coarse-grained zirconium polycrystals, where in each individual grain at least five mutually independent deformation systems are activated simultaneously. Additional accommodation deformation systems are observed in the area of grain boundaries. Otherwise, the deformation systems, at least up to 40% degree of deformation, are distributed homogeneously throughout the individual grain. This is explained by the marked tendency for twinning which extends throughout the whole grain.

At low deformation rates, in addition to twinning, only traces of the main slip system on prism planes in the *a* direction are found. Slip causes the lattice to rotate continuously but affects only relatively minor orientation changes, although the strain achieved can be great (Fig. 18).

Simultaneously with slip, twinning is activated even at low degrees of deformation, for example, 0.5%. In a polycrystal this takes place independently of the initial basal pole position even for orientations which are favorably oriented for prism slip and which in single-crystal experiments did not show any twinning even for high degrees of deformation [*123*]. However, a marked development of slip lines can be observed even at orientations that are less favorable for prism slip.

Twinning causes spontaneous lattice rotations. The angle of rotation and the shear strain of this inhomogeneous deformation process depend on the activated twinning system. Twinning can cause large lattice rotations, although the strain achieved is small. With increasing deformation, the volume twinned by the initially effective systems increases, essentially without new systems being activated.

From these examples, three areas for the activation of different twinning systems can be deduced, depending on the orientation. Correspondingly, the resulting lattice rotations are summarized schematically in Fig. 19 [*125*].

For basal positions in the area 0 to 50° from the Normal Direction (ND), irrespective of the azimuthal position of the basal pole, $\{11\bar{2}2\}$ twinning becomes preferentially operative. Under a compressive strain component

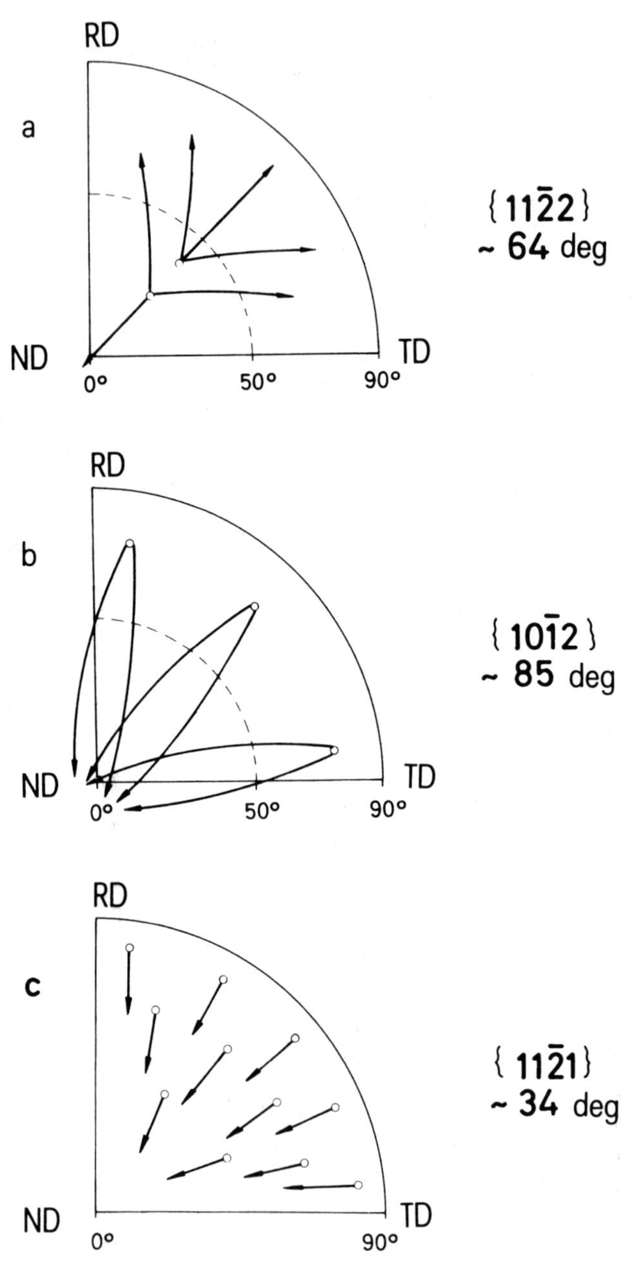

(*a*) Basal pole orientation 0 to 50° tilted from ND; lattice rotation by {11$\bar{2}$2} twinning.
(*b*) Basal pole orientation 50 to 90° tilted from ND; lattice rotation by {10$\bar{1}$2} twinning.
(*c*) Basal pole orientation 0 to 90° tilted from ND; accommodation lattice rotation by {11$\bar{2}$1} twinning.

FIG. 19—*Spontaneous lattice rotations of the basal pole due to twinning in zirconium (schematically, according to Ref* 125). *Depending on the starting orientation of the matrix, different twin systems are activated.*

parallel to the *c* axis, this twinning system rotates the basal pole by approximately 64° away from the center of the pole figure (Fig. 19*a*). By means of this rotation in the twinned structure, an orientation is usually achieved that is tilted 50 to 90° away from the ND. In extreme cases, however, the lattice can be rotated across the ND from a +50 towards a −15° tilt or vice versa.

For orientations in the area 50 to 90° from the ND, again irrespective of the azimuthal position of the basal pole, $\{10\bar{1}2\}$ twinning becomes preferentially operative. Under a tensile strain component parallel to the *c* axis, this twinning system rotates the basal pole by approximately 85° towards the center of the pole figure, also across the ND (Fig. 19*b*). This process is often activated as secondary twinning in orientations that were brought into this area (50 to 90° away from the ND) by $\{11\bar{2}2\}$ primary twinning.

The $\{11\bar{2}1\}$ twinning can be activated over the entire reference sphere. Under a tensile strain component parallel to the *c* axis, this twinning rotates the basal pole by approximately 34° preferentially towards the center of the pole figure (Fig. 19*c*). This twinning system, in accordance with the observations in Refs [*2,9,10*], becomes operative less often than the first two. Furthermore, the twins are restricted only to a small volume. It presumably acts only as a complementary system in order to accommodate strong lattice strains at grain boundaries or at twin interactions.

These twinning processes, described separately for better understanding, in reality interact with each other often as secondary or even higher-order twins [*125*]. In addition, the glide processes must also be taken into account, which results in a variety of complicated interactions of different deformation systems (see the example in Fig. 20).

The texture development starting from the basal pole position of the undeformed grains and ending with final stable texture, is given in Fig. 21. A prism pole figure taken after the last rolling step showed that a $[10\bar{1}0]$ direction was aligned parallel to the Rolling Direction (RD) as is commonly observed for cold deformation textures in zirconium.

The results show that, even at low deformations, twinning causes the basal poles to orient rapidly towards the ND, that is, parallel to the effective compressive force. At 20% total deformation, this process has already made great progress. From 20 to 40% deformation, the basal poles align further toward the final stable texture. At this point, the microstructure is completely twinned. Because, starting from a small grain size downwards (in this case, twin interfaces are equivalent to grain boundaries of crystallites), further twinning is prevented [*4*], the split of basal poles toward the TD final stable position of basal poles (0002) ± 20 to 40° in the Transverse Direction (TD) can be explained by means of pyramidal slip systems with a $(c + a)$ type Burgers vector on $\{11\bar{2}1\}$ or $\{10\bar{1}1\}$ planes. The existence of a $(c + a)$ type Burgers vector was confirmed in previous investigations

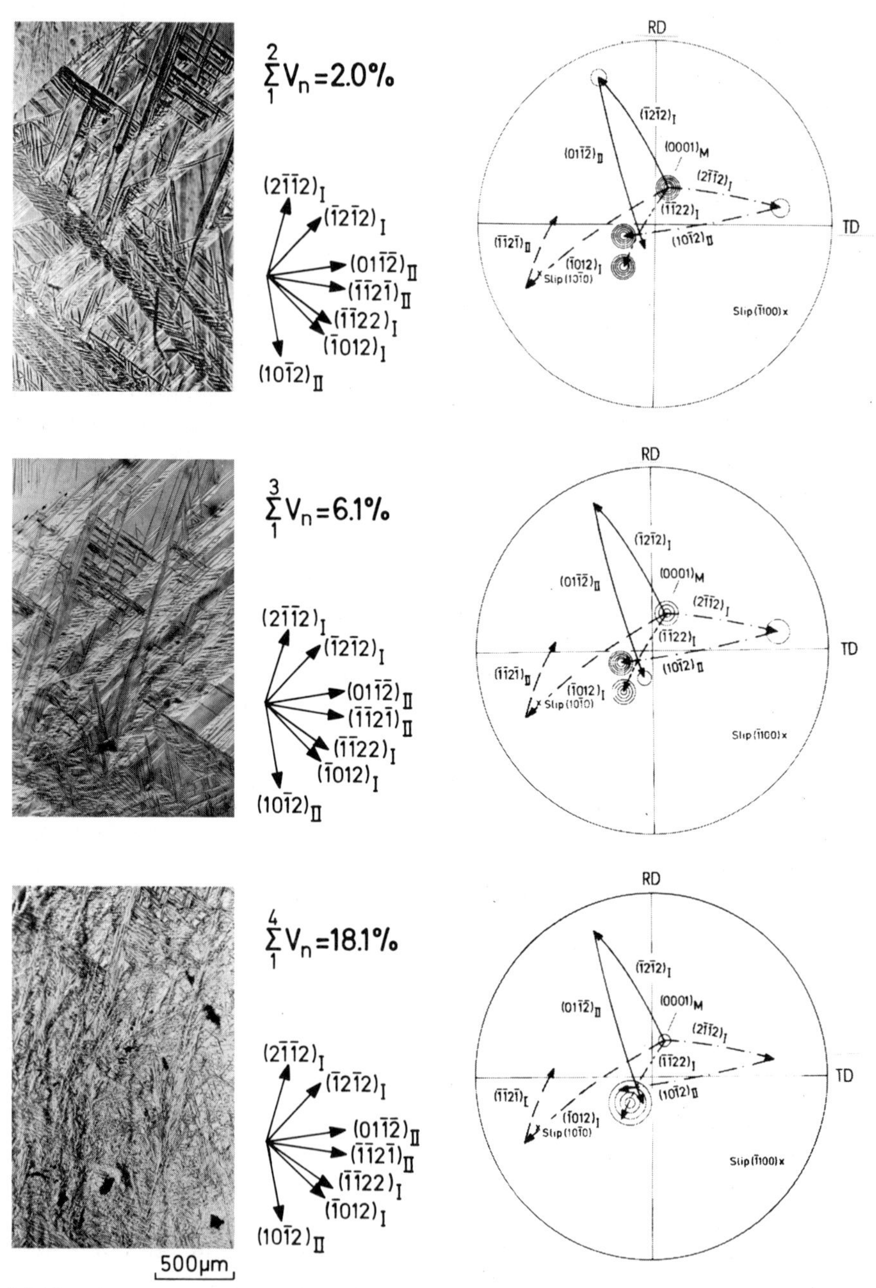

(*a*) The microstructure with the traces of the deformation systems (For the different degrees of deformation $\Sigma_1^i V_n$ a photograph was always taken of the same microsection).
(*b*) The associated degree of deformation and the indexed traces of the activated deformation systems.
(*c*) The stereographic projection with the corresponding by derived lattice rotations for the basal pole, derived from the different deformation systems. The measured texture intensities for the (0002) planes are also included, as indicated by the number and size of the circles.

FIG. 20—*Spontaneous lattice rotations due to twinning, Grain 18* (*according to Ref* 125).

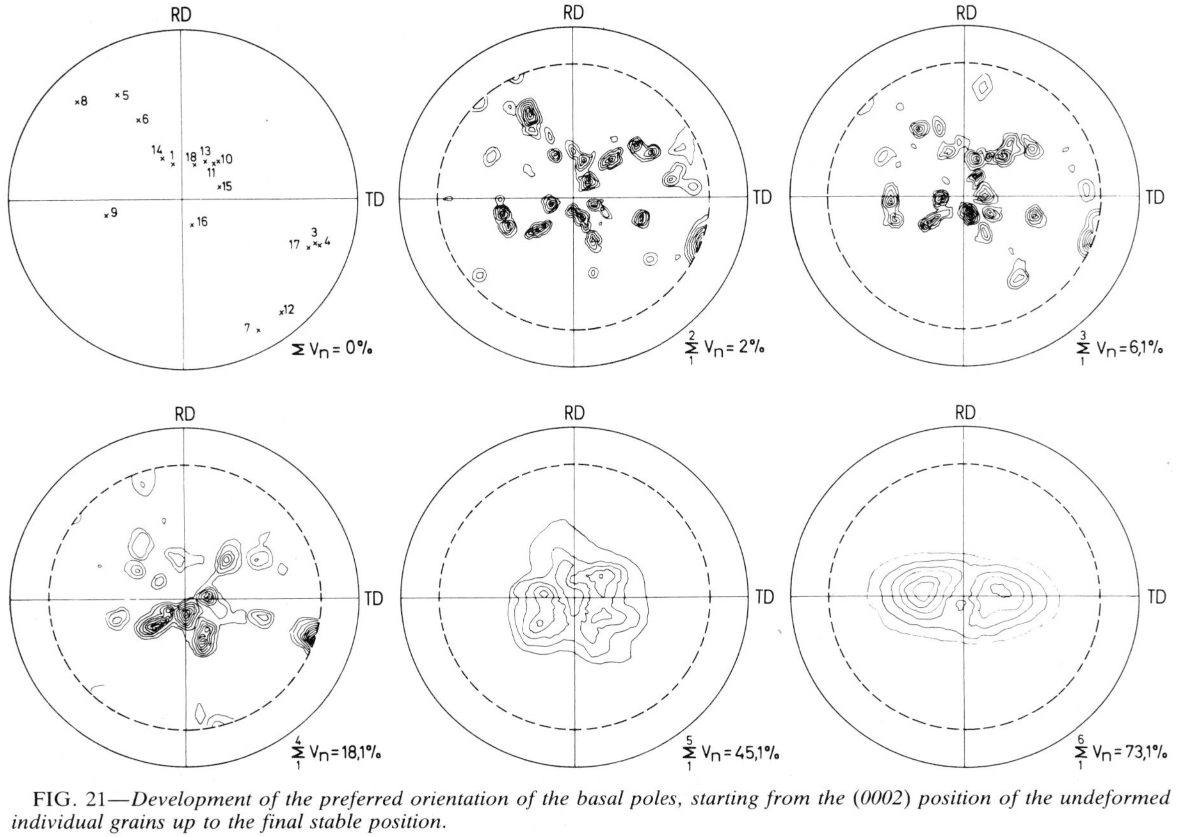

FIG. 21—*Development of the preferred orientation of the basal poles, starting from the (0002) position of the undeformed individual grains up to the final stable position.*

in single-crystal shear experiments, that is, under strong external restraint conditions [36]. Restraint conditions can also be caused by internal accommodation forces after high degrees of cold deformation in an intensely twinned microstructure.

Because of Schmid factor considerations out of the three possible pyramidal planes, $\{11\bar{2}1\}$, $\{10\bar{1}1\}$, and $\{11\bar{2}2\}$, the first mentioned is the most likely to render $(c + a)$ slip in zirconium [36]. As computer calculations for plate-rolling conditions show, the $\{11\bar{2}1\}$ pyramidal slip system leads to a maximum of the resulting Schmid factor, μ_{res}, for tilt angles of the basal pole between 20 to 40° toward TD [125], while the $\{10\bar{1}1\}$ system has decidedly smaller values in this interval. The $\{11\bar{2}2\}$ pyramidal slip system, which has been mentioned in this connection, shows the lowest values for this orientation range. Equivalent calculations for basal pole tilts toward RD show a sharp decline of the Schmid factor for all three systems after 10 to 15°, again in accordance with the texture measurements [125]. In addition, the $\{11\bar{2}2\}$ system offers, in contrast to the $\{11\bar{2}1\}$ and $\{10\bar{1}1\}$ systems, not twelve but only six slip systems to a family. Consequently for zirconium, the $\{11\bar{2}1\}$ pyramidal plane seems to be the most likely slip plane with a $(c + a)$ Burgers vector. Further investigations have to be made, to clarify the effect of $(c + a)$ slip on the stable final texture.

3.3 Characteristics of Deformation Textures in HCP Metals

As an introduction, a general review of the texture characteristics in hcp metals will be given. Particulars and detailed differences, possibly also due to older, less accurate measurements, are described at full length by Wassermann and Grewen [5], Dillamore and Roberts [6], Bunge [7] and Grewen [126]. In their work, the influences of impurities, alloy additions, deformation temperature, etc., are mentioned.

Most measurements are related to the basal pole distribution, which with respect to the anisotropy of hcp metals plays an important role. Less is known about the distribution of prism or pyramidal plane poles; the results are variable and do not give a completely uniform picture.

In all hcp metals for cold deformed semi-finished products (as for example, wire, sheet, or tubing), the final orientation of the basal plane (0001) is parallel to the direction of deformation. Deviations from this preferred orientation, that is, the tilts of the basal plane, exhibit characteristic differences, depending on the specific metal parameters [5–7,9,126]. Therefore, the textures of the hcp structures usually are categorized into three groups, according to the c/a axial ratio and thereby to the resulting operative deformation systems. As an example Fig. 22 shows schematically these three types of texture taking cold-rolled sheet textures.

For metals or alloys with an above-normal c/a ratio ($c/a > 1.633$), for example, zinc and cadmium, the position of the basal poles is tilted by ± 15

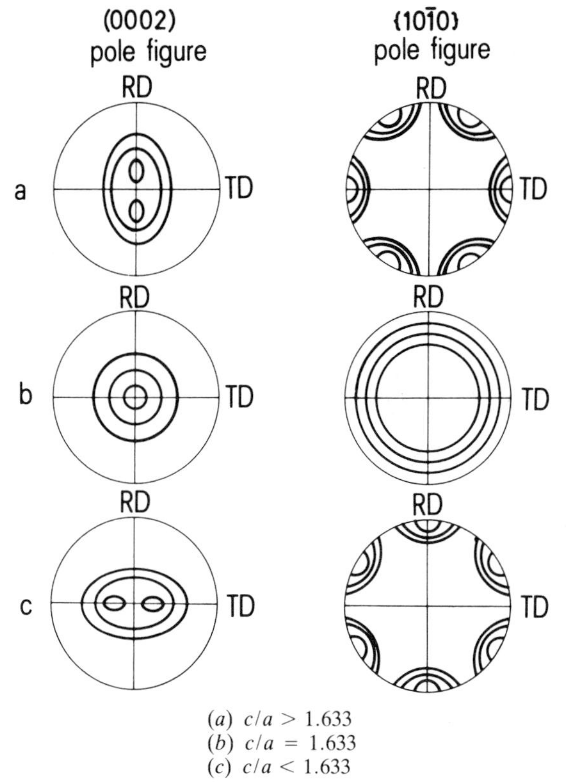

(a) $c/a > 1.633$
(b) $c/a = 1.633$
(c) $c/a < 1.633$

FIG. 22—*Sheet textures in hcp metals with different* c/a *ratios (schematically).*

to 25° from the normal direction toward the rolling direction. A [11$\bar{2}$0] direction is aligned parallel to the rolling direction (Fig. 22*a*).

For metals or alloys with a c/a ratio approximately equal to the ideal ratio ($c/a \cong 1.633$), for example, magnesium and cobalt, the basal poles concentrate in symmetry rotation around the normal direction. The [10$\bar{1}$0] and [11$\bar{2}$0] directions are randomly distributed around the pole of the basal plane ([0001] fiber texture) (Fig. 22*b*).

For metals or alloys with a below-normal c/a ratio ($c/a < 1.633$), for example, zirconium, titanium, and hafnium, the position of the basal poles is tilted by ±20 to 40° from the normal direction toward the transverse direction. A [10$\bar{1}$0] direction is aligned parallel to the rolling direction (Fig. 22*c*).

3.4 Deformation Textures in Zirconium and Zircaloy

The following is a review of the preferred crystallographic orientations that develop in zirconium and Zircaloy for the common deformation processes (wire drawing, sheet rolling, tube reduction) investigated.

In many texture analyses of wire, sheet, and tubing, it has been found that the basal planes always align parallel to the direction of the main deformation. The poles of the basal plane, however, orient themselves differently depending on the deformation conditions:

1. for wire drawing—random distribution in the radial-tangential plane (fiber texture) [5,*127*];
2. for sheet rolling—alignment in the normal direction, with a tendency to split toward the transverse direction by ±20 to 50° [5,*93*]; and
3. for tube reducing—alignment occurs in the radial-tangential plane. Depending on the reduction parameters, the alignment may occur in the radial direction, in the tangential direction in intermediate positions, or also randomly distributed in the radial-tangential plane [5,*128–139*].

For all cold deformed semi-finished products, a $[10\bar{1}0]$ direction is parallel to the direction of deformation.

In comparison to sheet rolling or wire drawing, the symmetry of the deformation process for tube reducing is lowest; that is, the degree of freedom in varying the deformation parameters is higher in tubing. The material flow caused by the reduction is strictly constrained by the deformations in the radial and tangential directions, resulting in an extension in the axial direction. In consequence, the conclusions about the operative forces, about the material flow, about the deformation mechanisms, and about the resulting textures are defined best. Therefore, the textures developed during tube reduction will be discussed first.

3.4.1 Texture in Tubing

The reduction of tubing is a complicated deformation process with a triaxial stress-strain condition that can vary according to different fabrication techniques. A characteristic of the deformation process is that reduction in cross-section, R_A, can be achieved by reduction in wall thickness, R_W, by reduction in diameter, R_D, or by any combination of R_W and R_D. The main deformations are the changes in wall thickness and in diameter that result in the axial extension of the tubing. The effective plastic deformations are primarily caused by compressive forces [*140*]. During rocking, for instance, the material will be compressed radially between the rolls and the mandrel and tangentially by the tapered sections of the rolls. The radial and tangential compressive forces can vary, however, according to the reductions in wall thickness and diameter. If the wall thickness reduction prevails, the compressive forces in the radial direction are stronger. If the diameter reduction prevails, the compressive forces in the tangential direction are stronger.

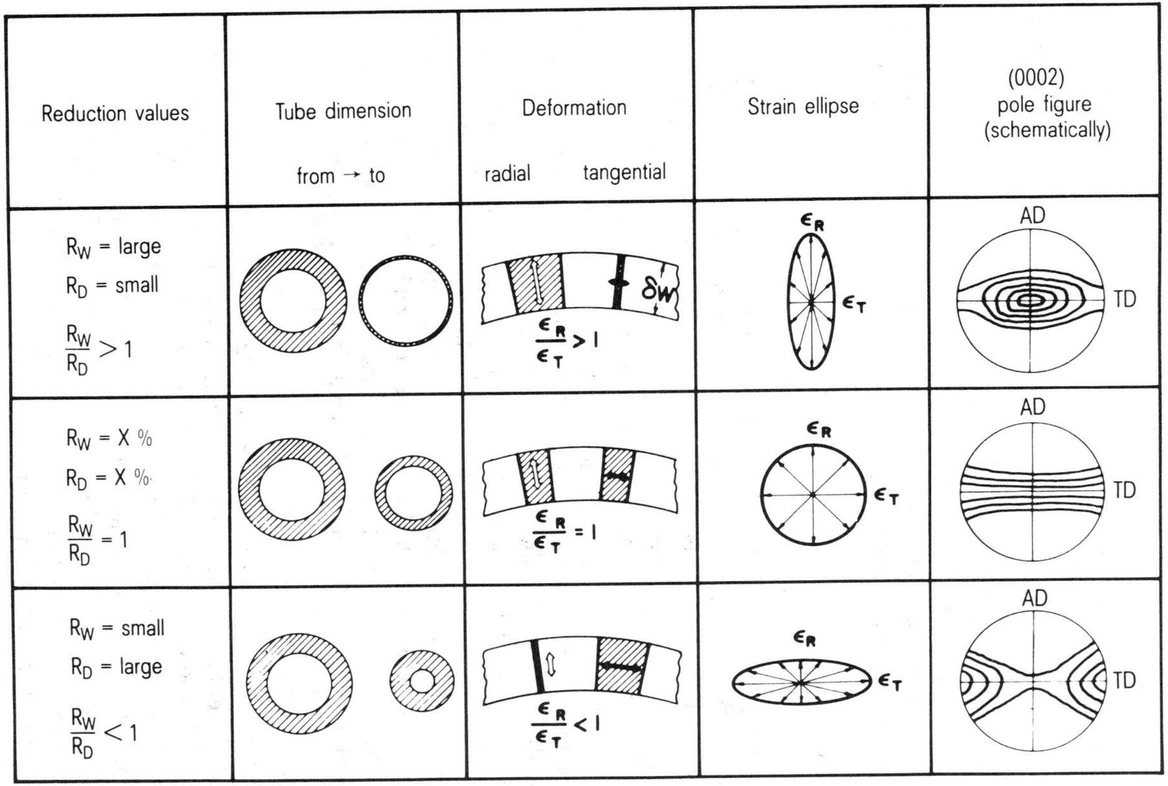

Reduction values	Tube dimension from → to	Deformation radial tangential	Strain ellipse	(0002) pole figure (schematically)
R_W = large R_D = small $\frac{R_W}{R_D} > 1$		$\frac{\epsilon_R}{\epsilon_T} > 1$ δW	ϵ_R ϵ_T	AD TD
R_W = X % R_D = X % $\frac{R_W}{R_D} = 1$		$\frac{\epsilon_R}{\epsilon_T} = 1$	ϵ_R ϵ_T	AD TD
R_W = small R_D = large $\frac{R_W}{R_D} < 1$		$\frac{\epsilon_R}{\epsilon_T} < 1$	ϵ_R ϵ_T	AD TD

FIG. 23—*Reduction values, deformation, resulting strain ellipse, and the derived basal pole figure (schematically), (according to Ref* 138).

3.4.1.1 Influence of the Relative Ratio of Reduction in Wall Thickness-to-Diameter—Norton and Hiller [*141*] first pointed out the influence of the wall thickness-to-diameter reduction on the texture development in cold-drawn carbon steel tubing. Systematic investigations on the influence of the reduction parameters (R_A, R_W, and R_D) on the texture development of Zircaloy tubing have shown that the factor controlling the texture is the ratio of wall thickness-to-diameter reduction [*138,139*]. Three extreme examples of the different reduction ratios, R_W/R_D, are shown on the ordinate of Fig. 23. The primary compressive deformations, the derived strain ellipse, and the resulting texture, are given on the abscissa [*138*]. For these three theoretical cases, it is assumed in the first instance that the material had a random orientation distribution before the respective deformation process. In addition, a very thin layer, δ_W, within the tube wall will be considered first.

Tubing deformed by a high wall thickness and a low diameter reduction ($R_W/R_D > 1$, Fig. 23*a*) is mainly compressed in the radial direction, whereas the compressive forces in the tangential direction are comparatively small. Since the basal poles primarily align parallel to the direction of the effective compressive force (see Section 2.2), the resulting strain ellipse leads to a (0002) figure with the basal poles preferentially parallel to the radial direction. If the wall thickness reduction is approximately equal to the diameter reduction ($R_W/R_D \cong 1$, Fig. 23*b*), that is, the deforming compressive forces operate with equal strength in the radial and the tangential directions, then the strain ellipse converts into a circle. The result is a fiber texture with a random distribution of basal poles in the radial-tangential plane. Tubing deformed by a small wall thickness and a large diameter reduction ($R_W/R_D < 1$, Fig. 23*c*) is mainly compressed in the tangential direction. Compared with the first case, the main deformation direction and consequently the resulting strain ellipse are tilted by 90°. Accordingly, in the (0002) figure, the basal poles align in the tangential direction.

3.4.1.2 Final Stable Positions—According to the ratio of the wall thickness-to-diameter reduction characteristic, final stable positions of the basal poles are found in Zircaloy tubing. Tubing deformed by a high wall thickness and a low diameter reduction ($R_W/R_D > 1$) leads to a texture with the normal of the basal plane primarily aligned parallel to the radial direction (within ±20 to 40° tilted to the tangential direction). If the wall thickness reduction is equal to the diameter reduction ($R_W/R_D = 1$), the result is a fiber texture with a random distribution of basal poles in the radial-tangential plane. Tubing deformed by a small wall thickness and a large diameter reduction ($R_W/R_D < 1$) leads to a texture with the poles of the basal plane primarily aligned parallel to the tangential direction [*138*].

3.4.1.3 Intensity Maxima Between the Stable Final Positions in the Radial and Tangential Directions—The existence of intensity maxima between the extreme positions in the radial and tangential directions can be explained if the texture existing in the material before the corresponding deformation process is taken into consideration. The direction in which the basal poles move depends on the ratio, R_W/R_D. How far the maxima shift and how strong their intensity is depend on the degree of deformation (that is, on the reduction in cross-section, R_A) and on the ratio, R_W/R_D [*138,139*].

3.4.1.4 Texture Gradient Through the Tube Wall—The concept of the operative forces, the resulting plastic deformations, and the derived texture development (see Fig. 23) must be applied to each single layer in the tube wall. That is, the ratio of wall thickness-to-diameter reduction must be taken into account for each layer, δ_W, through the wall of the tube, because both the radial and tangential deformations can change through the tube wall. Therefore, a texture gradient can occur. This development is explained in Fig. 24 in the example of a starting texture with a 45° split of basal poles toward the tangential direction; for simplicity it is assumed to be constant over the tube wall. From the outer to the inner diameter the spread of basal pole maxima can increase (Fig. 24*a*), it can decrease (Fig. 24*c*), or the position of the basal poles can remain constant (Fig. 24*b*).

The existence of these three forms of texture gradients is shown in Figs. 25 and 26 where the intensity distributions of (0002) poles in the radial-tangential plane are plotted for tubing with extreme reduction ratios [*137, 138*] and for tubing with deformations more commonly used in commercial tube reduction processes [*139*].

3.4.1.5 Texture Development in Sequential Steps—The dependence of texture development on the relative ratios of wall-to-diameter reduction can be followed continuously within one reduction pass or step-by-step in successive tube-reducing passes [*138,139*]. The texture of the material before the deformation step in question must always be taken into account. For tubes fabricated in several passes, an initial preponderance in diameter reduction is required so that the final dimensions of the tube can be achieved by a preponderance of reduction in wall thickness, if the basal poles are to be aligned preferentially in the radial direction. If the tangential position of the basal poles is desired, then the order of the wall thickness and diameter reduction has to be reversed. If the final stable position is reached, a further reduction in the same direction influences only the intensity and the sharpness of the texture but not the position of the maximum basal pole maxima [*139*].

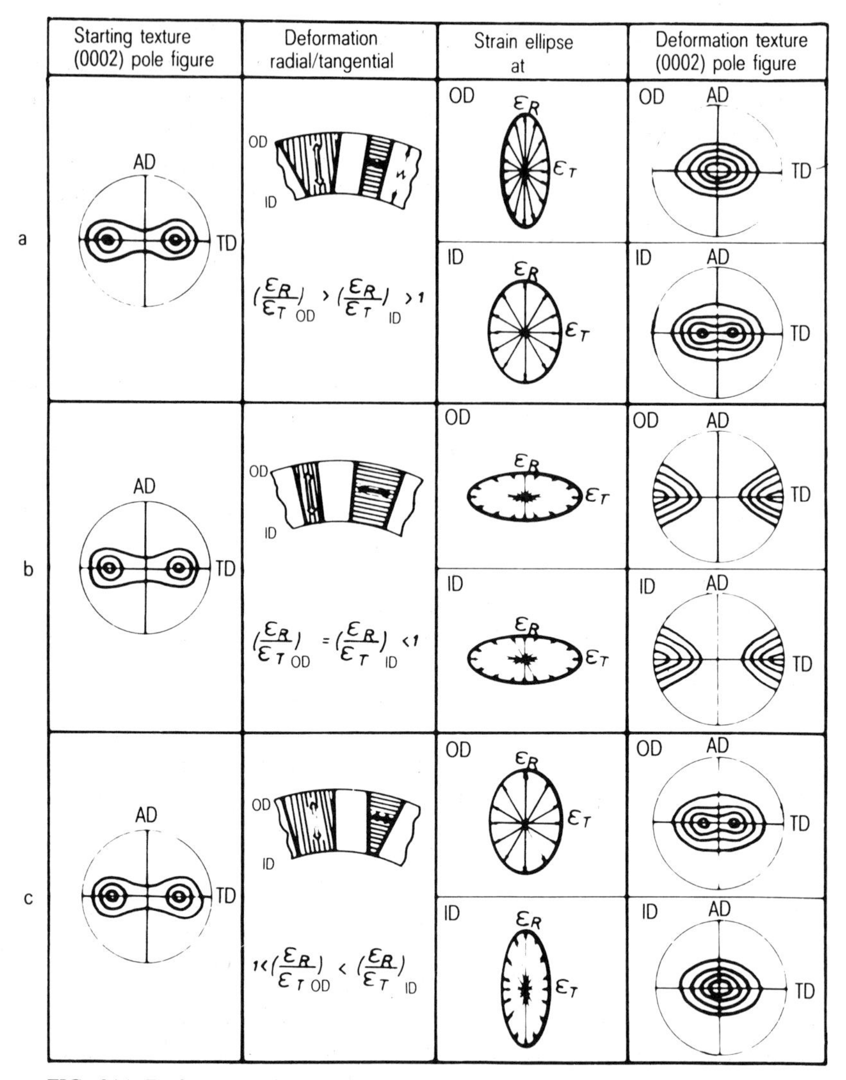

FIG. 24—*Explanation of the different types of texture gradients through the tube wall. The stress-strain concept of Fig. 22 is applied to each successive layer in the tube wall (according to Ref* 139).

3.4.1.6 Influence of the Fabrication Method—Neither fabrication method (tube reducing or rocking, mandrel drawing, planetary ball swaging, hammer swaging) nor a change in the deformation direction causes significantly different textures when the processes are carried out with corresponding reductions in wall thickness, diameter, and area [*138,139*]. Knowing these dependences, it is possible to tailor the texture of Zircaloy tubing within

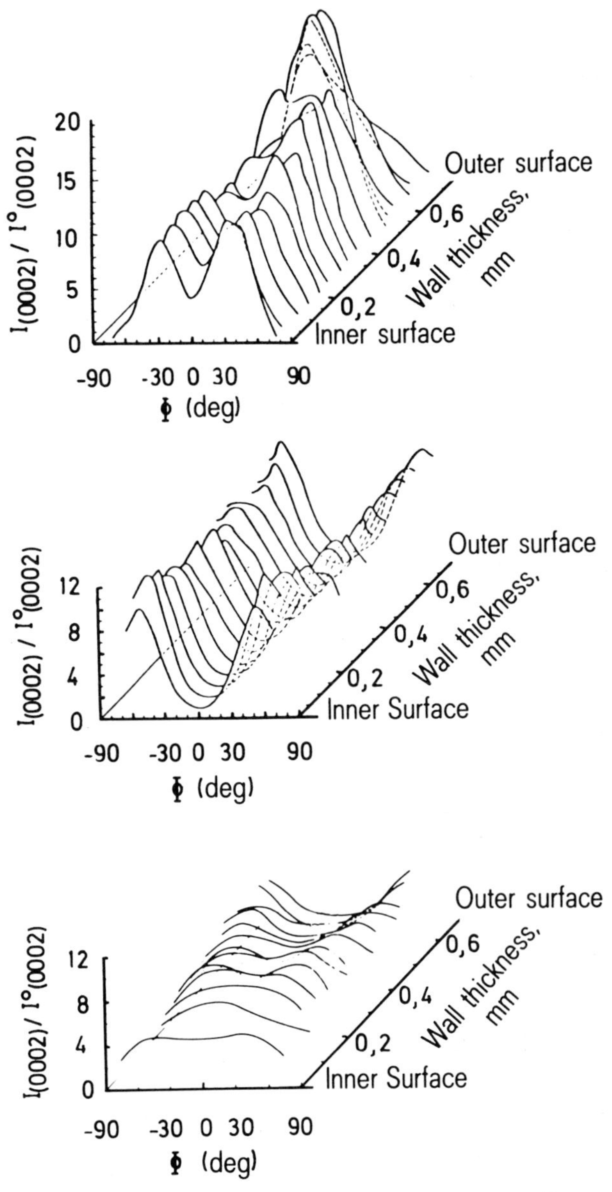

(*a*) Spread of basal pole maxima decreases from inner to outer diameter.
(*b*) Spread of basal pole maxima stays constant from inner to outer diameter.
(*c*) Spread of basal pole maxima increases from inner to outer diameter.

FIG. 25—*Intensity distribution of (0002) basal poles in the radial-tangential plane through the wall thickness of tubes with extreme reduction values (according to Ref* 138), *where* Φ *is the angle between the normal to the sample and the normal to the diffracting net plane*).

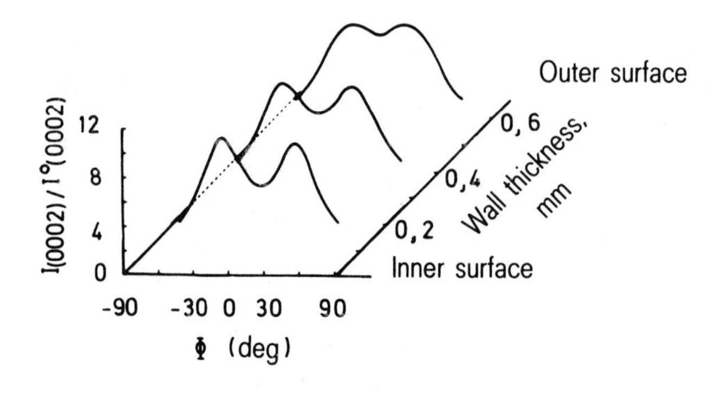

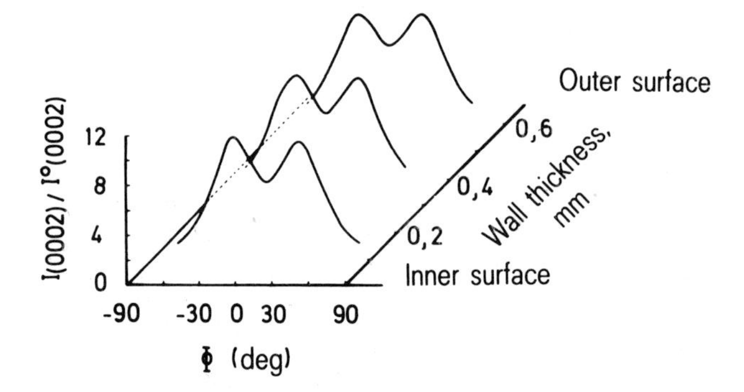

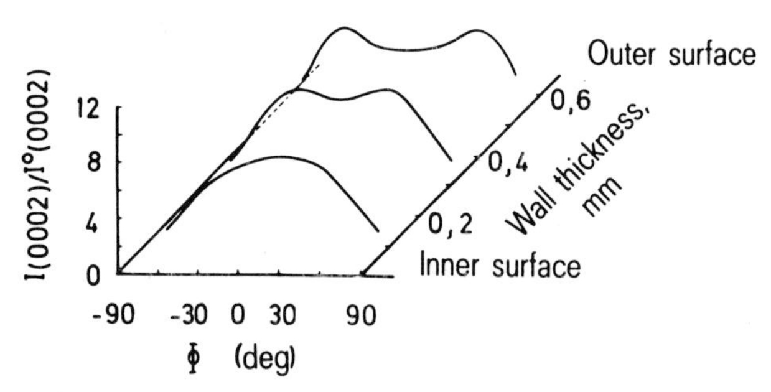

(*a*) Spread of basal pole maxima decreases from inner to outer diameter.
(*b*) Spread of basal pole maxima stays constant from inner to outer diameter.
(*c*) Spread of basal pole maxima increases from inner to outer diameter.

FIG. 26—*Intensity distribution of basal poles in the radial-tangential plane through the wall thickness of tubes with reductions more commonly used in commercial tube fabrication processes* (*according to Ref* 139), *where* Φ *is the angle between the normal to the sample surface and the normal to the diffracting plane*).

the given limits. The stress-strain concept for the texture formation, derived from the deformation process of tube reducing (Fig. 27), applies also for textures developed during sheet rolling or wire drawing.

3.4.2 Textures in Sheet

During the rolling process, the sheet is compressed in the normal direction, a process by which the sheet is lengthened in the rolling direction, whereas it is barely in the transverse direction. (This is usually true for a ratio of sheet-width/thickness > 6). In the plane normal to the rolling direction, the strain ellipse is characterized by a compressive deformation in the normal direction and a small compressive deformation in the transverse direction. Since in cold-deformed sheet the basal poles always tend to align parallel to the deforming compressive force, the basal poles are preferentially parallel to the normal direction with a tendency to spread by ±20 to 40° toward the transverse direction.

The texture shown in Fig. 27*b* is identical to the texture in tubing for $R_W/R_D > 1$ as shown in Fig. 23*a* and 27a_1. This is to be expected, since the stress-strain conditions are comparable.

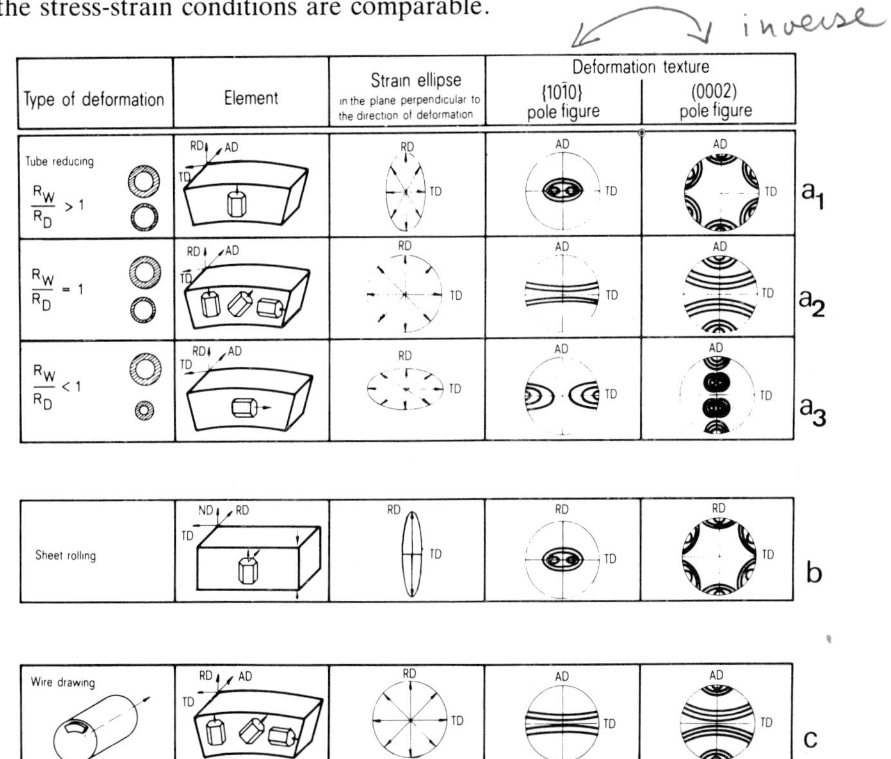

FIG. 27—*Application of the stress-strain concept to texture development for zirconium and Zircaloy in tubing* (a_1 *to* a_3), *sheet* (b), *and wire* (c).

3.4.3 Textures in Wire

During the wire drawing process, the material is compressed concentrically, resulting in a diameter decrease and finally an axial extension of the wire. In the plane normal to the axial direction, the strain ellipse converts into a circle (Fig. 27*c*). Consequently, because of the dependence of the compressive direction (see Section 3.2), the basal poles become randomly distributed in the radial-tangential plane [*19,127,142*].

This fiber texture (Fig. 27*c*) is identical to tubing texture, if the tubing is deformed with $R_W/R_D \cong 1$ (Figs. 23*b* and 27a_2). This is to be expected, since the stress-strain conditions are comparable. In wire drawing, these deformation conditions are valid for each volume element of the cross section. One can visualize the wire as consisting of concentric tubes of different diameters. Under the assumption that the volume stays constant during the deformation process, the visualized tubes of different diameters must all be deformed with $R_W/R_D = 1$; otherwise they would separate or overlap.

3.5 Annealing Textures in Zirconium and Zircaloy

The annealing textures in zirconium and Zircaloy can be summarized as follows [*143*] (a more detailed review of texture development in these materials and in hexagonal metals in general is given in Refs *5* and *6*).

The basal pole figure does not change significantly during annealing. In part, it can be observed that the basal poles tend to concentrate more toward the normal or radial direction, respectively; in part, the pole density is simultaneously somewhat increased or decreased. In contrast, the prism $\{10\bar{1}0\}$ pole figure is more revealing for the observation of the development of the annealing textures. During recrystallization with increasing annealing temperature, the basal planes rotate continuously by $\pm 30°$ around their pole, so that in the final stable position, instead of a $[10\bar{1}0]$ direction (cold deformation texture), a $[11\bar{2}0]$ direction becomes parallel to the rolling or axial direction, respectively.

The annealing textures in tubing appear to be more complex than those of rolled and annealed sheet. In tubing, different features of annealing textures can develop, depending on the relative reductions in wall thickness and diameter, the degree of deformation, the temperature of deformation, and on the deformation method before annealing (extruding, rocking, drawing) [*143*].

Chapter 4—Mechanical Anisotropy

The objective of metallurgical texture investigation is not only to elucidate the interactions between deformation mechanisms and texture development (Chapter 3) but also to explain their influence on the mechanical anisotropy of textured polycrystals. If one knows the interactions between deformation mechanisms and texture development, it should then be possible to predict the mechanical anisotropy of a component with a given texture by means of the deformation mechanisms.

The previously-mentioned limitations (Section 3.2*a* to *f*) to applying the deformation theories to the texture development of hcp metals are certainly also valid in the discussion of the interactions between deformation mechanisms and mechanical anisotropy.

The deformation mechanisms of hcp metals, in general, and zirconium, in particular, discussed in Chapter 2 illustrated their pronounced anisotropy. The factors influencing the deformation mechanism have been discussed in Section 2.3; in particular, for hcp metals, the rather complicated correlation between single-crystal properties and the behavior of the textured polycrystal were treated in Sections 2.3.4 to 2.3.6.

4.1 Influence of the Deformation Mechanisms on the Mechanical Anisotropy of Textured Zircaloy Material

4.1.1 Zirconium and Zircaloy Plates

The order of magnitude of the anisotropy in a Zircaloy-4 plate with a typical rolling texture ((0002) ±20 to 40° in a transverse direction, $\langle 10\bar{1}0\rangle$) can be seen in the following example. If tensile and compressive samples are taken from the plate in such a way that in one case the basal poles are preferentially aligned parallel to the direction of the applied force and in the other preferentially perpendicular to it, the values listed in Table 8 may be obtained. Depending on the orientation and the direction of force, the ratio of the yield strengths is approximately 3:2:1 in this example.

This is due to the various deformation systems that are for deformation at room temperatures in Case a primarily $\{11\bar{2}2\}$ twinning, in Case b primarily $\{10\bar{1}2\}$ and $\{11\bar{2}1\}$ twinning, and in Case c primarily $\{10\bar{1}0\}$ $\langle 11\bar{2}0\rangle$ prism slip, if one employs single-crystal behavior as a basis (see Chapter 2, especially Section 2.3.2, Table 7).

This anisotropic effect does not necessarily restrict the range of industrial application; anisotropy may be an advantage if, for example, the direction

TABLE 8—*Yield strength according to the direction and type of loading in a textured Zircaloy-4 plate* (*values according to Ref* 93).

Case	Loading	Yield Strength, N/mm²
a	compressive stress ∥ *c* axis	$\sigma_{0.2}$ = 2150
b	tensile stress ∥ *c* axis	$\sigma_{0.2}$ = 1430
c	tensile or compressive stress ⊥ *c* axis	$\sigma_{0.2}$ = 650

of extreme loading coincides with the direction of highest strength ducts preferred orientation. On the other hand, it is possible, for deformation procedures such as rolling, bending, etc., where high ductility is required, to select specifically the direction in which the material can be deformed easily due to its directionality.

4.1.2 Zircaloy Tubing

The influence of the texture on the mechanical anisotropy as well as the causal interactions with the deformation mechanisms will be discussed using the example of a Zircaloy tube. Zircaloy tubes are used as fuel element cladding materials for pressurized and boiling water reactor systems. It is important to know which texture provides the best mechanical properties for Zircaloy cladding tubes under reactor conditions. For example, exploiting a "texture-hardening effect" one could, by controlling the texture while maintaining the safety margin, make the cladding tubes thinner, thus improving the neutron yield.

The texture of Zircaloy tubing can be controlled by the choice of reduction sequence, especially by the ratio of wall thickness-to-diameter reduction that has been described in Chapter 3. The importance of these textures for the strength properties of tubing and the way in which their behavior can be explained on the basis of the knowledge of the deformation mechanisms will first be discussed with the aid of idealized loading conditions for both possible extreme positions of the basal poles in Zircaloy tubing, that is, in radial and tangential directions.

If cladding tubes are machined out of a sufficiently thick Zircaloy plate with pronounced rolling texture so that the tube axis corresponds to the rolling direction, then continuously changing preferred orientations of the basal poles (with the extreme positions possible in Zircaloy tubing) are obtained at Positions I and II around the circumference of the tube [*91*] (Fig. 28). The texture at Position I corresponds to the tube texture with preferential position of the basal poles in the radial direction. The texture at Position II is rotated through 90° compared to that of Position I; that is, the basal poles lie preferentially in the tangential direction.

Thus, it is possible to measure the purely textural influence on the mechanical anisotropy of Zircaloy tubing, since other parameters such as the

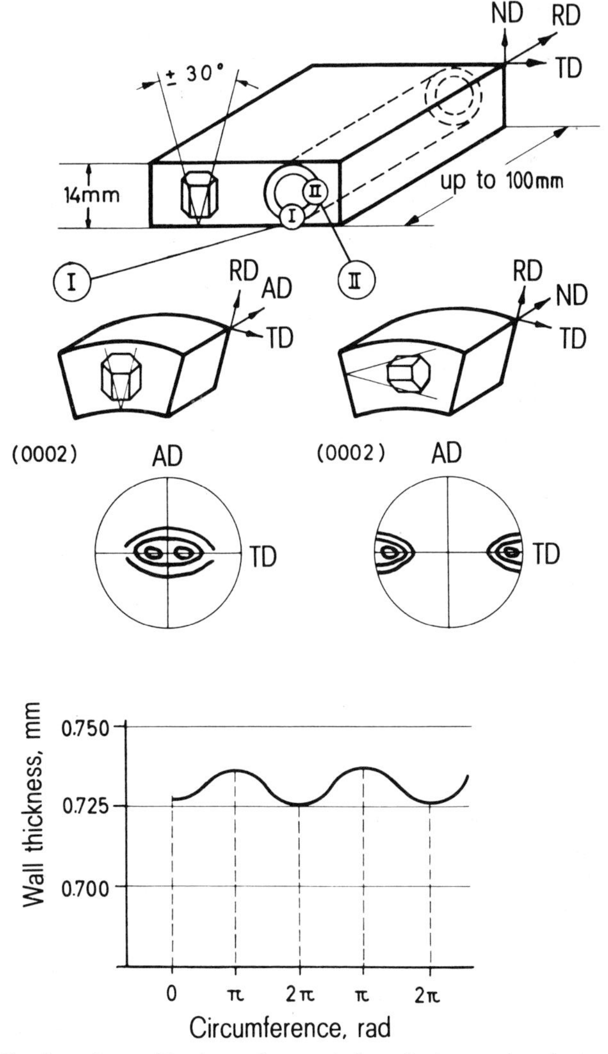

FIG. 28—*Zircaloy tube machined out of textured plate, single-crystal analogies, (0002) pole figures in the tube at Positions I and II, and wall thickness changes around the circumference (according to Ref* 91).

degree of cold deformation, of recrystallization, etc., are necessarily constant [*91*].

The texture of the test tube can be roughly described by the single-crystal analogies sketched in Fig. 28. The more pronounced the texture the more the properties of the polycrystal resemble those of the single crystal (with the restrictions referred to in Section 2.3.5). Thus, in order to predict the

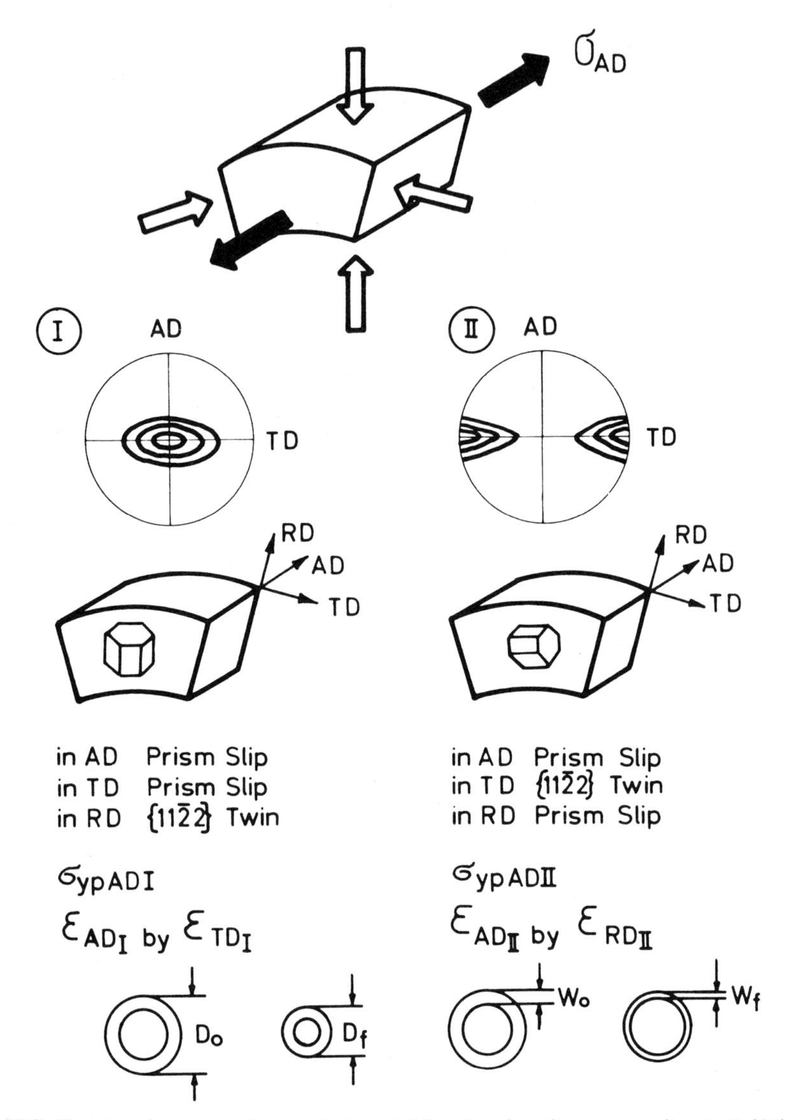

FIG. 29—*Zircaloy tube under tension in axial direction* (*prediction according to Ref* 91).

mechanical anisotropy, the operative deformation systems known from a single crystal experiment can be employed.

For a first estimation of the mechanical anisotropy, it is sufficient to differentiate between the activation of slip or twinning modes. This differentiation is primarily dependent on the crystallographic orientation of the *c* axis with respect to the direction of force, as the comparison of Schmid factors has shown (see Section 2.3.1). This explains why it is sufficient to take the basal pole figure into account for a first approximation of the mechanical anisotropy. The example of a tube under tensile stress in the axial direction will show this in more detail.

The upper part of Fig. 29 shows a tube element that is stressed axially (solid arrows) while contractile strains result in the radial and tangential directions (outlined arrows). Therefore, for a tube with basal poles preferentially aligned in the radial direction (Case I), the stress-strain conditions would activate the following deformation mechanisms according to Section 2.1 to 2.3:

1. Prism slip by tension in the axial direction.
2. Again prism slip by transverse contraction in the tangential direction.
3. $\{11\bar{2}2\}$ twinning by transverse contraction in the radial direction.

Since the critical resolved shear stress for prism slip is lower than the corresponding stress initiating $\{11\bar{2}2\}$ twinning (see Section 2.3.2, Table 7), the tube will elongate primarily by reduction of the diameter with little change in wall thickness.

For a tube element with preferred basal pole positions in the tangential direction (Case II), the stress-strain conditions would cause:

1. $\{10\bar{1}0\}$ $\langle 11\bar{2}0\rangle$ prism slip by tension in the axial direction.
2. $\{11\bar{2}2\}$ twinning by transverse contraction in the tangential direction.
3. Prism slip by transverse contraction in the radial direction.

Because of the differences in critical resolved shear stresses, the tube in this case would be expected to elongate by decreasing its wall thickness with little change in diameter (Fig. 29, Case II).

The strength (for example, the yield stress) would be equally low for either type of texture, because prism slip is operative in the axial direction in both cases.

The deformations just predicted for the two types of texture can be followed by a single test on the tube machined out of the sheet (Fig. 30). During the tension test, the tube became elliptical, that is, Position I, the curvature decreased while the wall thickness stayed nearly constant; on the other hand, at Position II, the tube wall decreased in thickness while the curvature stayed nearly constant. The difference in reductions in wall thickness can be also seen in the two axial microsections of Positions I and II. Additionally, Position I exhibits a 90° fracture while Position II has a 45° fracture. The two types of fractures also differ in the fractographic microstructure in that the former exhibits pure tensile dimples and the latter shows shear dimples. These results, obtained in tests at room temperature and 400°C, are in agreement with tension tests on round specimens

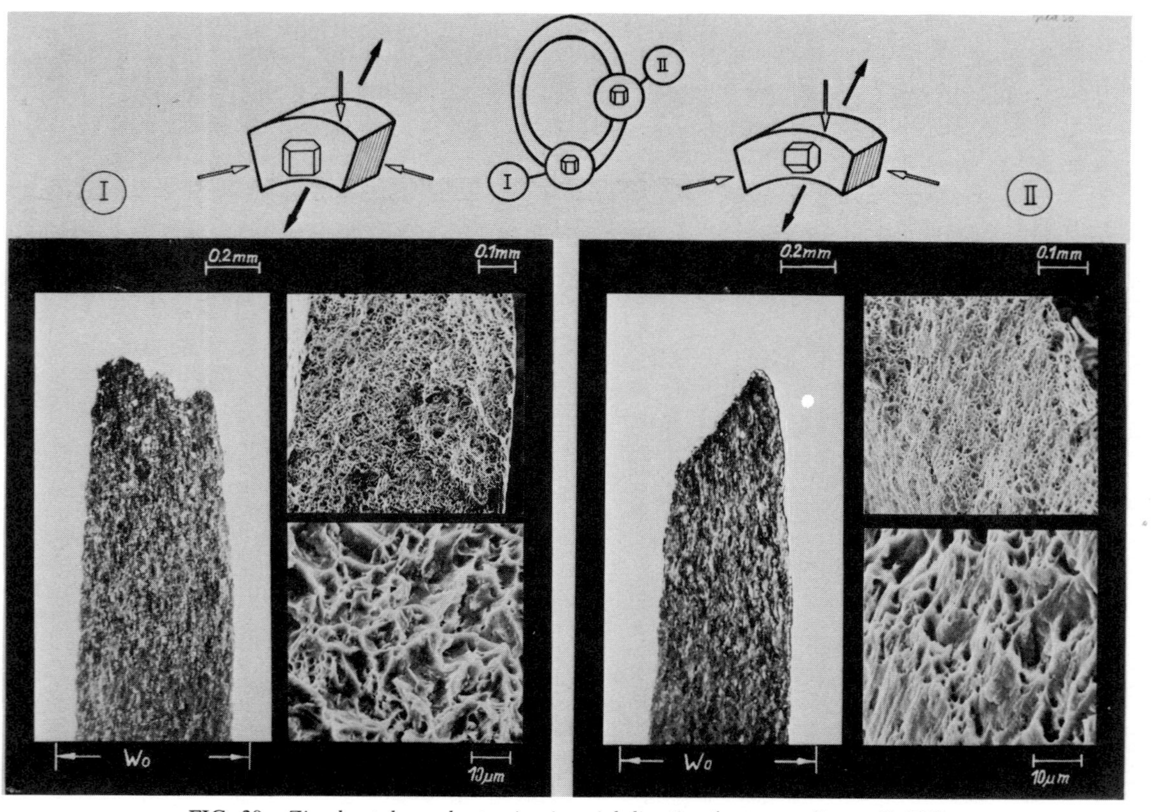

FIG. 30—*Zircaloy tube under tension in axial direction (test according to Ref* 91).

cut out of a textured sheet [*140*], and on tubing with basal pole variations around the circumference [*144*].

This relatively simple type of loading shows that the analysis of operative deformation mechanisms allows a qualitative estimation of the mechanical anisotropy of textured components. Further estimations of the mechanical anisotropy and their experimental verification have been performed for a tube under internal pressure, axial bending moment, axial torsion and superposed bending moment, and external pressure (collapse) [*91*]. It can be stated that, in all these cases, not only do the experimental results from the tube taken from the textured Zircaloy plate agree with the theoretically derived behavior, but also the results described in the literature [*144–147*].

4.2 Representation of Mechanical Anisotropy for Biaxial Stress Conditions

4.2.1 Yield Loci

In addition to establishing a correlation between the properties of single crystals and the behavior of textured polycrystalline materials with the aid of deformation theories (see Section 2.3), it is worthwhile to be able to estimate the behavior of an anisotropic material for particular multiaxial stress conditions.

To solve the complicated relationships encountered, one usually starts with two-dimensional stress conditions, especially since in many of the technical load conditions one of the three stress components is zero or negligibly small. The general von Mises criterion (see Eq 18, Section 2.3.4) can, assuming that

$$\sigma_3 = 0 \tag{20}$$

be reduced to the following

$$\sigma_1^2 + \sigma_2^2 - \sigma_1 \sigma_2 = K_f^2 \tag{21}$$

This elliptical equation provides a two-dimensional representation of the stress conditions whose transgression leads to yielding. It is therefore termed the yield locus.

Figure 31 shows a comparison of the von Mises yield locus [*99*] with a further yield locus by Tresca [*148*]. The maximum distortion criterion of von Mises accounts for the total stress conditions, whereas the maximum shearing stress criterion of Tresca only allows for a part of the stress components present in the material, that is, the smallest and largest principal stress corresponding to Mohr's circle for plane stress

$$\sigma_1 - \sigma_2 = K_f \tag{22}$$

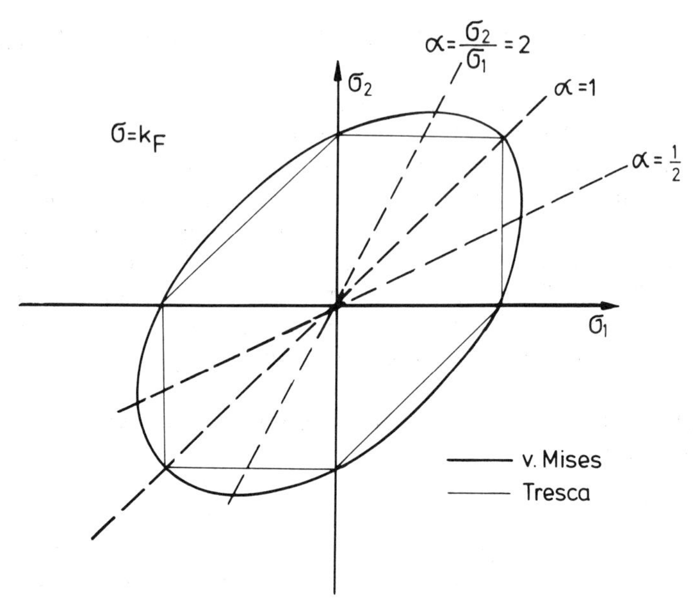

FIG. 31—*Huber-Mises-Henky yield criterion (ellipse) and Tresca yield criterion (hexagon) (according to Ref* 148).

This explains the differences (0 to about 12%) from the von Mises boundary curve.

The inherent anisotropy of metallic materials is not allowed for in these initial considerations. Correspondingly, there are differences between real, measured yield loci and those according to von Mises and Tresca. To render these deviations amenable to mathematical analysis, Hill [*100*] proposed anisotropic factors based solely on computational, rather than crystallographic grounds. For two-dimensional conditions, the distortion behavior in the indicated directions, for example, in a metal sheet, are correspondingly described by the so-called r coefficient. This r coefficient is defined as the ratio of the logarithmic distortions in the width and thickness directions of a uniaxial sample subjected to tensile stress, which has been taken from a metal sheet in various directions, i.

$$r_i = \frac{\ln b/b_0}{\ln d/d_0} \tag{23}$$

whereby b_0 and d_0 are the initial width and thickness of the flat bar, and b and d are the width and thickness, respectively, after plastic deformation.

The r coefficient is a measure of the resistance of the material to contraction in its thickness direction. It can be interpreted in three cases as follows:

1. $r < 1$, that is, the material contracts more in thickness than in width.
2. $r = 1$, that is, the material exhibits mechanically isotropic behavior.
3. $r > 1$, that is, the material contracts largely in its width.

Figures 32*a* and *b* show the influence of the *r* coefficient on the yield locus shape in the cases of sheet material without surface anisotropy or with surface and thickness anisotropy [*8*]. Because the propositions made by Hill are based purely on computational anisotropy factors, only radially symmetric yield loci can be constructed; possible dependences on the sign of the yield stress are not included.

Initial attempts at providing a crystallographic basis for the mechanical anisotropy with the aid of selection criteria for effective deformation mechanisms, were made by Taylor [*101*] using the principle of the minimum internal work performed in deformation as well as by Bishop and Hill using the equivalent principle of the maximum external work performed in deformation. A clear prediction of the mechanical anisotropy for triaxial stress conditions has, however, yet to be achieved. Using the simplifying assumptions of uniaxial and biaxial loading conditions, single crystals were initially investigated and the methods of calculation then applied to the single-crystal condition of the textured material. Hereby, the difficulties increase with the number of possible deformation systems from the fcc via the bcc to the hcp lattice structure with its marked interaction of slip and twinning, each caused by different systems [*8,103–110*].

In the context of these suggested solutions, Piehler and Backofen [*103*] described the complete solution of the real yield locus on a crystallographic basis for fcc and bcc materials. The real yield locus lies between two separately derived yield loci, which are termed the principal stress yield locus (within the real yield locus) and the principal strain yield locus (outside the real yield locus). Both yield loci have two normal stresses as the orthogonal reference system, which are so chosen that the third normal stress is zero. If the principal directions of stress and strain coincide, then the two yield loci are identical. Figure 33 shows these yield loci, derived to allow for the crystallographic deformation systems, in comparison with the von Mises ellipse, using the example of the orientation (110) [$1\bar{1}2$]. An experimental investigation of sheet copper crystals for this and other ideal orientations exhibited good agreement [*149*].

For hcp metals, the theory was expanded by Chin and Mammel [*107*] as well as by Thornburg and Piehler [*108*] that corresponded to the principle of maximum external work and was experimentally checked by Hosford [*109*]. Based on these propositions, Dressler, Matucha, and Wincierz [*94,110*]

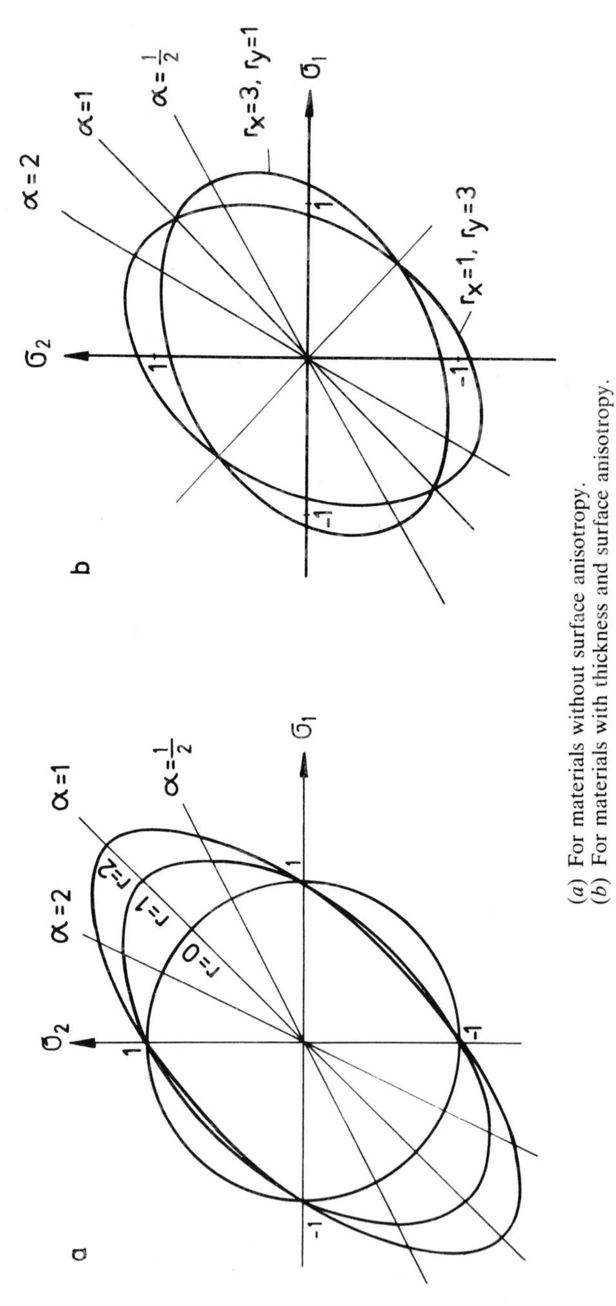

(*a*) For materials without surface anisotropy.
(*b*) For materials with thickness and surface anisotropy.

FIG. 32—*Yield loci for anisotropic yielding in sheet thickness (according to Ref* 8).

examined Zircaloy cladding tubes with regard to their yield behavior and assigned the experimentally obtained yield loci to the measured textures. The results, thus derived, are shown in Fig. 34. Starting from the ideal tilt of the basal poles, where $\gamma = 0$ to 30° and $\gamma = 70°$ toward the tangential direction, the yield loci (theoretically derived allowing for the various slip and twinning systems as well as the adapted critical shear stress conditions) are compared with the experimental yield loci. Also here one can observe the good agreement between theory and experiment.

4.2.2 Creep Loci

In relation to the yield loci found for short-time loading, for long-time loading creep loci can also be determined [*150*]. Figure 35 shows such creep loci for four differently textured Zircaloy tubes. The creep loci show, as in the case of the yield loci, the influence of the anisotropy on their shape, as well as the suitability of the material for certain loading conditions.

The creep loci represent locus diagrams of constant comparative creep rate, $\dot{\epsilon}_g$

$$\dot{\epsilon}_g = \frac{\dot{\epsilon}_1\sigma_1 + \dot{\epsilon}_2\sigma_2}{\sigma_g} \tag{24}$$

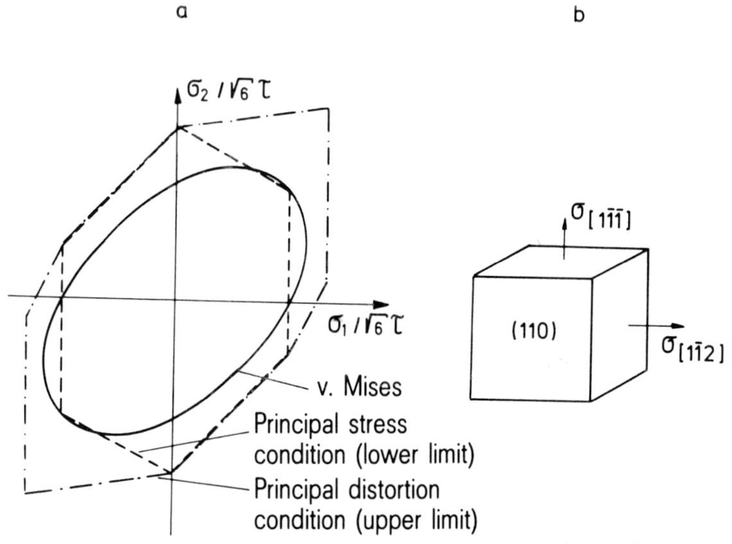

FIG. 33—*Comparison between theoretically calculated yield loci* (a) *according to von Mises as well as Piehler and Hosford for the orientation given in Part* b (*according to Ref* 103).

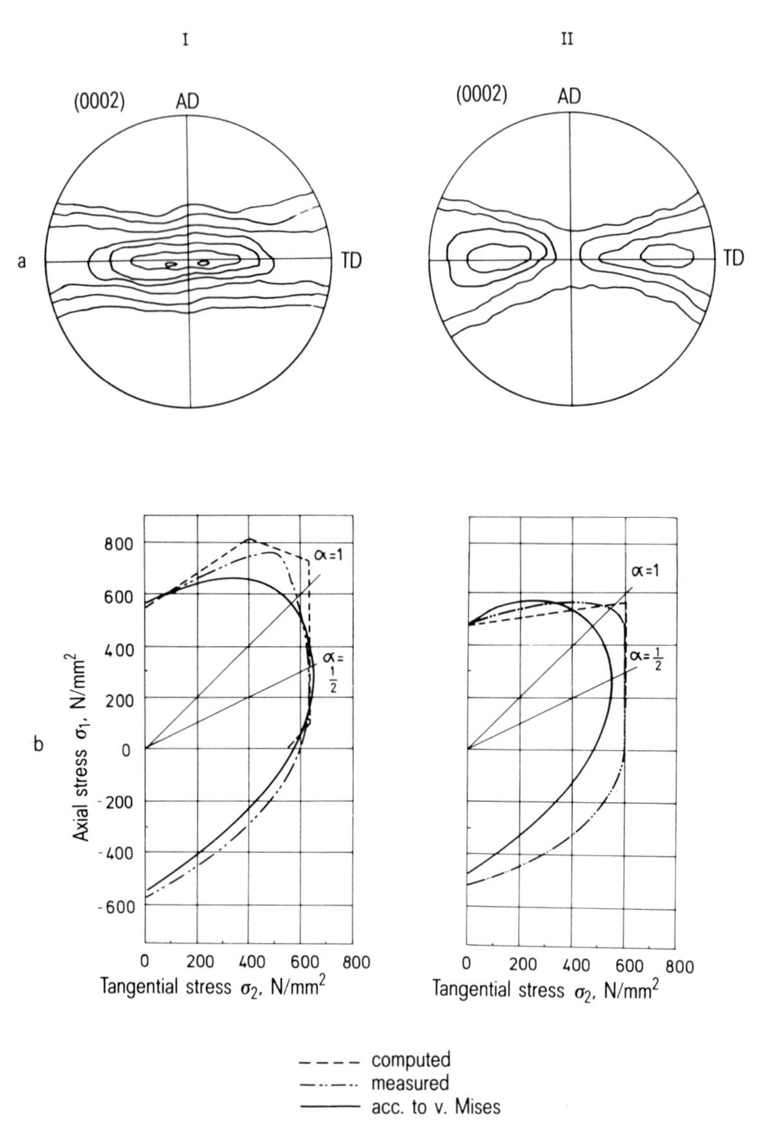

FIG. 34—*Yield loci* (b_I *and* b_{II}) *of Zircaloy-4 cladding tubes with various basal pole textures* (a_I *and* a_{II}) *for:* (*I*) *basal pole inclination in TD = approximately ± 0 to 30°; and* (*II*) *basal pole inclination in TD = approximately ± 70°* (*according to Ref* 110).

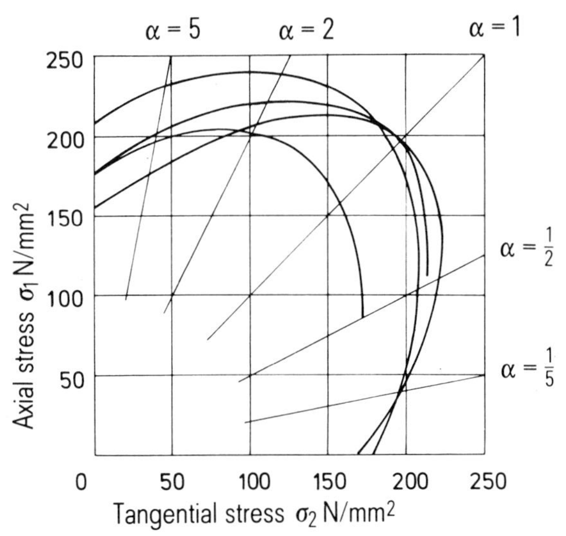

FIG. 35—*Creep loci for various Zircaloy tube textures, obtained at 400°C and a constant deformation efficiency of* W $= 3 \cdot 10^{-2}$ *N/mm²h (according to Ref* 150).

whereby σ_g represents the simplification of Hill's propositions by Holicky and Schroeder [*151*]

$$\sigma_g = \sqrt{\sigma_1^2 + A_2\sigma_2^2 + 2A_{12}\sigma_1\sigma_2} \tag{25}$$

The locus could be derived by obtaining pairs of values, σ_1 and σ_2, in axial and tangential directions, respectively, for which a constant ϵ_g occurred. To determine the locus experimentally under these conditions is, however, difficult. Therefore, the creep rates, ϵ_1 and ϵ_2, were measured for the various stress conditions and stresses. From this, the work performed in deformation can be calculated

$$W = \dot{\epsilon}_g \cdot \sigma_g \tag{26}$$

and interpolation can be made between these values [*150*].

An assignment of the anisotropy factors calculated from crystallographic considerations to the experimentally determined anisotropy factors on the basis of the short-time deformation mechanisms has, however, proved to be unsatisfactory.

4.2.3 Burst Loci

A further method of assessing the plastic behavior of metallic materials is offered by the so-called burst loci [*150,152,153*]. The burst locus is, analogous to the yield locus, the two-dimensional representation of the stress conditions whose transgression leads to burst (Fig. 36). The elliptical shape of these curves is less marked than that of the yield loci. In addition, they cannot be described using the propositions made by Hill. An attempt was made by Duncomb [*152*] to link them to mechanical processes in the metal, although a satisfactory solution has yet to be achieved.

4.3 Effects of Creep and Irradiation on the Operative Deformation Systems in Zircaloy Tubing under Reactor Conditions

By analyzing the operative deformation mechanisms, it is possible to find selection criteria for desirable textures in Zircaloy cladding tubes. However, a prerequisite for this is that the rather complex loading conditions acting on the cladding tube should be ascertained and their influence be correctly assessed.

For Zircaloy tubing under reactor service conditions, the loadings are very complex. Various types of stresses can be applied simultaneously and can alter in magnitude, direction, and rate of change with time and temperature. In addition to short-term deformation, long-term deformation (creep) as well as cyclic deformation and irradiation effects can play important roles.

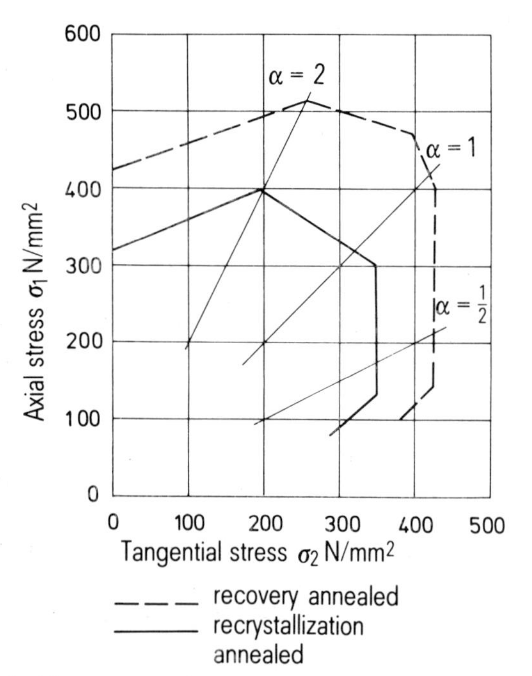

FIG. 36—*Burst loci for recovery and recrystallization annealed Zircaloy tubes. Rate of stress increase 200 to 400 N/mm²min at 400°C (according to Ref* 150).

The effects of creep and irradiation on the deformation systems operative in zirconium and Zircaloy are summarized in the following sections.

4.3.1 Creep

At lower loading rates and with increasing temperature, dislocation reactions and diffusion controlled processes (for example, cutting, climb, cross-slip, jog motion of screw dislocations, Cottrell unlocking, and grain boundary sliding) are favored over martensitic transformation type deformations [*154–158*], which also allow plastic deformations.

4.3.2 Neutron Irradiation

Neutron irradiation produces point defects that can cluster, segregate to dislocations, and finally create forest obstacles. Through the resulting dislocation climb and thermal cutting processes, irradiation influences slip behavior, although twinning should not be affected appreciably [*154*].

For short-time deformations, an increase in strength and a decrease in ductility is observed [*154,159–161*]. For long-time deformations, the effect of neutron irradiation is a pronounced increase in creep rate, stress relaxation, and rupture ductility [*150,154,162,163*]. In the temperature range from 300 to 400°C, recovery of the irradiation damage occurs; at the higher limit, irradiation damage is nearly completely annealed out as fast as it occurs [*164,165*].

Although the detailed mechanisms for creep [*155–158*] and creep during irradiation [*166–168*] are different, in both cases, deformations with a c-component may be accomplished by slip with a $(c + a)$ type Burgers vector and has been shown to become operative at elevated temperatures under restraint [*36*]. For creep under reactor service conditions, pyramidal slip may also become operative in Zircaloy tubing. Wavy slip lines observed on post-irradiation-test creep specimens [*155*] can be interpreted as cross slip pyramidal planes. A further confirmation of the existence of $(c + a)$ slip can be ascertained from the findings of Bell [*169*], who observed slip traces in deformed, irradiated Zircaloy by electron microscopy, which indicate $(c + a)$ slip.

The different *crss* on the newly activated slip systems may lead to anisotropic behavior that is characteristic of creep, or creep under neutron irradiation, and is different from the short-term mechanical behavior. The large variation in parameters affecting creep, or creep under irradiation, explains the difficulty in finding consistent models for the activated deformation mechanisms. Further investigations, specifically directed to creep deformation mechanisms under neutron irradiation, are needed.

Chapter 5—Summary

The deformation systems in hexagonal close-packed (hcp) metals are not as numerous and not as symmetrically distributed as in cubic ones. Thus in plastic deformation, twinning competes with slip and may, depending on the deformation conditions, play an essential role. In addition to its effect on the transformation behavior and the corresponding stacking fault energy, the c/a axial ratio of the hcp structure also determines which deformation mechanisms are activated. This axial ratio varies from one metal to the other and can reach values both larger and smaller than those for ideal sphere packing. This trait prevents the deformation behavior of hcp metals from being considered en bloc, in contrast to fcc and bcc metals. In order to explain the conditions in zirconium, the well-established relationships of hcp metals are given, and their dependences on the metal-specific parameters of the hexagonal structure are discussed. The interactions between deformation mechanisms and texture formation on the one side and deformation mechanisms and mechanical anisotropy on the other can be likewise transferred to other hcp metals, if one takes into account the differences in dependence of the metal-specific parameters.

In the α structure of zirconium, slip is activated on prism planes in the a direction within the temperature range up to 500°C. In the same direction at elevated temperatures also basal slip and in regions of stress concentration slip on $\{10\bar{1}1\}$ planes have been observed at elevated temperatures. Furthermore, pyramidal slip in a $(c + a)$ direction has been observed under restraint conditions and at elevated deformation temperatures. Apart from the pyramidal slip system deformations with a c component are normally explained by twinning. Under tensile stresses in a c direction, primarily $\{10\bar{1}2\}$ twins and sometimes $\{11\bar{2}1\}$ twins are activated. Under compressive loading in a c direction, $\{11\bar{2}2\}$ twinning and $\{10\bar{1}1\}$ twinning at elevated temperatures are observed. In some cases, the less well-defined $\{11\bar{2}3\}$ twinning mode has been also reported.

The effects influencing the deformation mechanisms are discussed generally for hcp metals and particularly for zirconium and Zircaloy. For single crystals under uniaxial loadings, the Schmid factor, the critical resolved shear stress, and the direction of deformation are decisive for the activation of the deformation systems. In addition, for polycrystals under multiaxial loadings, the accommodation conditions and the influence of the texture must be taken into account.

The low offer of slip systems, with their asymmetrical distribution as well as the strict crystallographic orientation twinning relationships, cause in hcp metals the formation of a strong deformation texture. This is, however, characteristically conditioned by the metal-specific parameters of the hexagonal structure. In zirconium and Zircaloy, the development of a marked deformation texture is caused by the complicated interaction between slip and twinning. By virtue of twinning, even small deformation rates lead to large lattice rotations, which change the orientation of the crystallites where all basal poles align in the direction of the compressive force. The fact that the preferred orientation, which is spread in the transverse direction in zirconium and Zircaloy, is also retained as the final stable position at elevated temperatures is explained by $(c + a)$ pyramidal slip. The decisive factor in texture development is the material flow, whose degree of freedom is low for seamless tube reduction, whereby it determines the reduction in cross-section, wall thickness R_W, and diameter R_D. This permits more precise prediction of the operative forces and the resulting deformation mechanisms.

The experiences gained with tubing can be transferred to sheet rolling and wire drawing. The decisive factor for the development of tube textures is the relative ratio of wall thickness-to-diameter reduction, R_W/R_D. For $R_W/R_D > 1$, the basal poles are preferentially aligned in the radial direction. For $R_W/R_D = 1$, the basal poles are randomly distributed in the radial-tangential plane. For $R_W/R_D < 1$, the basal poles are preferentially aligned in the tangential direction.

The sheet texture is identical to the tube texture for $R_W/R_D > 1$. In both examples, the material flow is characterized by a preponderance of wall thickness reduction.

The fiber texture of wires is identical to the texture of tubes for $R_W/R_D = 1$. One can visualize the wire deformation as corresponding to that of concentric tubes with different diameters to comply with tube reduction rates $R_W/R_D = 1$ under the condition of constant volume. Independent of the fabrication method for the cold-worked semifinished products just mentioned, a $[10\bar{1}0]$ direction always aligns itself parallel to the direction of elongation. When the dependence of the texture development on the reduction parameters is known, it is possible to tailor the texture of Zircaloy tubing in the context of the described extreme alignments, so they can be optimally matched to the requirements.

On the other hand, the deformation mechanisms also account for the pronounced mechanical anisotropy of a textured material. This is discussed using the example of a sheet and a tube of textured Zircaloy under uniaxial loading. The agreement between theoretical prediction and actual experimental behavior also applies to much more complicated loading conditions. For biaxial loading conditions, the anisotropic behavior is normally illustrated by yield loci, creep loci, or burst loci according to the criterion employed, that is yield stress, creep rate, or fracture stress, respectively. Depending on the texture and the loading conditions, an attempt is made to correlate the shape of the loci to the operative deformation mechanism. In this way, it is possible, for instance, to find selection criteria for the desirable texture in Zircaloy cladding tubes.

For an exact estimation, the complex, load-dependent conditions during operation, together with the variously activated long-time or short-time (creep) deformation mechanisms (with or without irradiation), must also be taken into account.

Chapter 6—References

[1] Hall, E. O., *Twinning and Diffusionless Transformations in Metals,* Butterworths, London, 1954.
[2] *Deformation Twinning,* R. E. Reed-Hill, J. P. Rosi, and H. C. Rogers, Eds., Gordon and Breach, Science Publishers, New York and London, Vol. 25, 1964.
[3] Christian, J. W., *Theory of Transformations in Metals and Alloys,* Pergamon Press, Oxford, 1965.
[4] Partridge, P. G., *Metallurgical Reviews,* Vol. 12, No. 118, 1967, p. 169.
[5] Wassermann, G. and Grewen, J., *Texturen Metallischer Werkstoffe,* Springer, Berlin, 1962.
[6] Dillamore, I. L. and Roberts, W. T., *Metallurgical Reviews,* Vol. 10, 1965, p. 271.
[7] Bunge, H. J., *Kristall und Technik,* Vol. 6, 1971, p. 667.
[8] Hosford, W. S., Jr., and Backofen, W. A. in *Fundamentals of Deformation Processing,* 9th Sagamore Conference, Army Material Research Conference, Syracuse University Press, 1964, p. 259.
[9] Picklesimer, M. L., *Electrochemical Technology,* Vol. 4, 1966, p. 289.
[10] Reed-Hill, R. E., *Review of High Temperature Materials,* Vol. 1, 1972, p. 97.
[11] *Metals Handbook,* 8th ed., T. Lyman, Ed., American Society for Metals, Novelty, Ohio, Vol. 1, 1961.
[12] Barret, C. S. and Massalski, T.B., *Structure of Metals,* McGraw-Hill, New York, 1966.
[13] Smallman, R. E., *Modern Physical Metallurgy,* Butterworths, London, 1970.
[14] Weertman, J. and Weertman, J. R., *Elementary Dislocation Theory,* MacMillan, London, 1964.
[15] Seeger, A., *Moderne Probleme der Metallphysik,* Springer, Berlin, 1965.
[16] Schmid, E. and Boas, W., *Kristallplastizität,* Springer, Berlin, 1936.
[17] Peierls, R., *Proceedings,* Physics Society, Vol. 52, 1940, p. 34.
[18] Nabarro, F. R. N., *Proceedings,* Physics Society, Vol. 58, 1947, p. 669.
[19] Sagel, K. and Zwicker, U., *Zeitschrift fuer Metallkunde,* Vol. 46, 1955, p. 835.
[20] Thompson, N., *Proceedings,* Physics Society, London, Vol. B66, 1953, p. 481.
[21] Berghezan, A., Fourdeux, A., and Amelinckx, S., *Acta Metallurgica,* Vol. 9, 1961, p. 464.
[22] Schottky, G., Seeger, A., and Speidel, V., *Physica Status Solidi,* Vol. 9, 1965, p. 231.
[23] Thornton, P. R. and Hirsch, P. B., *Philosophical Magazine,* Vol. 3, 1958, p. 738.
[24] Harris, J. R. and Masters, B. C., *Proceedings,* Royal Society, Vol. A292, 1966, p. 240.
[25] Conrad, H. and Perlmutter, I., AFML-TR-65-310, Air Force Materials Laboratory, Wright-Patterson AFB, Ohio, 1965.
[26] Devlin, J. F., *Journal of Physics, F: Metal of Physics,* Vol. 4, 1974, p. 1865.
[27] Friedel, J., *Electron Microscopy and Strength of Crystals,* Interscience Publishers, New York and London, 1963.
[28] Akhtar, A. and Teghtsoonian, A., *Acta Metallurgica,* Vol. 19, 1971, p. 655.
[29] Akhtar, A. and Teghtsoonian, A. *Metallurgical Transactions,* Vol. 6A, 1975, p. 2201.
[30] Aldinger, F. and Jönsson, S. in *Beryllium 1977,* 4th International Conference on Beryllium, Royal Society, London, 1977, p. 4/1–10.
[31] Rosenbaum, H. S. in *Deformation Twinning,* R. E. Reed-Hill, J. P. Rosi, and H. C. Rogers, Eds., Gordon and Breach Science Publishers, New York and London, Vol. 25, 1964, p. 43.
[32] Tyson, W., *Acta Metallurgica,* Vol. 15, 1967, p. 574.
[33] Frank, F. C. and Thompson, N., *Acta Metallurgica,* Vol. 3, 1955, p. 30.
[34] Regnier, P. and Dupcuy, J. M., *Physica Status Solidi,* Vol. 39, 1970, p. 79.
[35] Aldinger, F. and Petzow, G., *Gefüge und Bruch, Materialkundl Technische Reihe,* Vol. 3, Berlin, Stuttgart, Gebrüder Borntraeger, 1977.
[36] Tenckhoff, E., *Zeitschrift fuer Metallkunde,* Vol. 63, 1972, p. 192.

[37] Damjano, V. V., London, G. J., and Conrad, H., *Transactions,* The Metallurgical Society, American Institute of Mining, Metallurgical and Petroleum Engineers, Vol. 242, 1968, p. 987.
[38] Kronberg, M. L., *Acta Metallurgica,* Vol. 9, 1961, p. 970.
[39] Kronberg, M. L., *Journal of Nuclear Materials,* Vol. 1, 1959, p. 85.
[40] Reed-Hill, R. E. in *Deformation Twinning,* R. E. Reed-Hill, J. P. Rosi, and H. C. Rogers, Eds., Gordon and Breach Science Publishers, New York and London, Vol. 25, 1964, p. 295.
[41] Stoloff, N. S. and Gensamer, M., *Transactions,* The Metallurgical Society, American Institute of Mining, Metallurgical and Petroleum Engineers, Vol. 227, 1963, p. 70.
[42] Westlake, D. G. in *Deformation Twinning,* R. E. Reed-Hill, J. P. Rosi, and H. C. Rogers, Eds., Gordon and Breach Science Publishers, New York and London, Vol. 25, 1964, p. 29.
[43] Chyung, C. K. and Wei, C. T., *Philosophical Magazine,* Vol. 15, 1967, p. 161.
[44] Balsdale, K. C. A. and King, R., *Physica Status Solidi,* Vol. 10, 1965, p. 175.
[45] Balsdale, K. C. A., King, R., and Puttik, K. E., *Physica Status Solidi,* Vol. 18, 1966, p. 491.
[46] Hauser, F. E., Landon, P. R., and Dorn, J. E., *Transactions,* American Society for Metals, Vol. 48, 1956, p. 986.
[47] Orowan, E., *Nature* (London), Vol. 149, 1942, p. 643.
[48] Rosi, F. D., *Transactions,* American Institute of Mining, Metallurgical and Petroleum Engineers, Vol. 200, 1954, p. 58.
[49] Rosi, F. D., Perkins, F. C., and Seigle, L. L., *Transactions,* American Institute of Mining, Metallurgical and Petroleum Engineers, Vol. 206, 1956, p. 115.
[50] Churchman, A. T., *Acta Metallurgica,* Vol. 3, 1955, p. 22.
[51] Reed-Hill, R. E. and Martin, J. L., *Fifth Quarterly Progress Report,* AEC Contract No. AT(38-1)-252, Atomic Energy Commission, Jan. 1963.
[52] Reed-Hill, R. E. and Martin, J. L., *Sixth Quarterly Progress Report,* AEC Contract No. AT(38-1)-252, Atomic Energy Commission, March 1963.
[53] Cahn, R. W., *Journal,* Institute of Metals, Vol. 79, 1951, p. 129.
[54] Honeycombe, R. W. K., *Journal,* Institute of Metals, Vol. 80, 1952, p. 45.
[55] Cahn, R. W., *Acta Metallurgica,* Vol. 1, 1953, p. 49.
[56] Chen, N. K. and Maddin, R., *Transactions,* American Institute of Mining, Metallurgical and Petroleum Engineers, Vol. 189, 1951, p. 531.
[57] Rosenbaum, H. S., *Acta Metallurgica,* Vol. 9, 1961, p. 742.
[58] Partridge, P. G. and Roberts, E., *Acta Metallurgica,* Vol. 12, 1964, p. 1205.
[59] Roberts, E. and Partridge, P. G., *Acta Metallurgica,* Vol. 14, 1966, p. 513.
[60] Rapperport, E. J., *Acta Metallurgica,* Vol. 7, 1959, p. 254.
[61] Rapperport, E. J., "Room Temperature Deformation Processes in Zirconium," NMI-1199, Technical Information Service Extension, Oak Ridge, TN, 1958.
[62] Sokurskii, I. N. and Protsenko, L. N., *Soviet Journal of Atomic Energy,* Vol. 4, 1958, p. 579 (English translation).
[63] Rapperport, E. J. and Hartley, C. S., "Deformation Modes of Zirconium at 77 K, 300 K, 575 K and 1075 K," NMI-1221, Nuclear Metals, Inc., Concord, MA, 1959.
[64] Howe, L. M., Whitton, J. L., and McGurn, J. F., *Acta Metallurgica,* Vol. 10, 1962, p. 773.
[65] Pollard, J., Rzepski, Mme., and Lehr, P. in *9 Colloque de Metallurgie, Etude sur la Corrosion et la Protection du Zirconium et de des Alliages,* M. Salese and M. Chaudron, Eds., Press Universitaires de France, Paris, 1966, p. 248.
[66] Baldwin, D. H. and Reed-Hill, R. E., *Transactions,* American Institute of Mining, Metallurgical and Petroleum Engineers, Vol. 223, 1965, p. 248.
[67] Bailey, J. E., *Journal of Nuclear Materials,* Vol. 7, 1962, p. 300.
[68] Martin, J. L. and Reed-Hill, R. E., *Transactions,* American Institute of Mining, Metallurgical and Petroleum Engineers, Vol. 230, 1964, p. 780.
[69] Rosi, F. D., personal communication to A. R. Kaufman, cited in; Rapperport, E. J., *Acta Metallurgica,* Vol. 14, 1966, p. 513.
[70] Jensen, J. A. and Backofen, W. A., *Canadian Metallurgical Quarterly,* Vol. 11, 1972, p. 39.

[*71*] Akhtar, A., *Journal of Nuclear Materials,* Vol. 47, 1973, p. 79.
[*72*] van Bueren, H. G., *Imperfections in Crystals,* North-Holland Publishing Co., Amsterdam, 1961,
[*73*] Roberts, C. S., *Magnesium and Its Alloys,* Wiley, New York and London, 1960.
[*74*] Tegart, W. J. McG., *Philosophical Magazine,* Vol. 9, 1964, p. 339.
[*75*] Chalmers, B., *Physical Metallurgy,* Wiley, New York and London, 1959.
[*76*] Williams, D. N. and Eppelsheimer, D. S., *Journal,* Institute of Metals, Vol. 81, 1952–53, p. 553.
[*77*] Hirsch, P. B., Howie, A., Nicholson, R. B., and Pashley, V. W., *Electron Microscopy of Thin Crystals,* Butterworths, Washington, DC, 1965.
[*78*] Heimendahl, M. v., *Einführung in die Elektronenmikroskopie,* Werkstoffkunde Bd. 1, Vieweg, Braunschweig, 1970.
[*79*] Reed-Hill, R. E., "An Evaluation of the Role of Deformation Twinning in the Plastic Deformation of Zirconium," *Ninth Quarterly Report,* AEC Contract No. AT(38-1)-252, Atomic Energy Commission, Nov. 1963.
[*80*] Warren, M. R. and Beevers, C. J., *Metallurgical Transactions,* Vol. 1, 1970, p. 1657.
[*81*] Tenckhoff, E., *Zeitschrift fuer Metallkunde,* Vol. 63, 1972, p. 729.
[*82*] Westlake, D. G., *Acta Metallurgica,* Vol. 9, 1961, p. 327.
[*83*] Partridge, P. G. and Roberts, E. in *Proceedings,* 3rd European Regional Conference on Electron Microscopy, Prague, 1964, p. 213.
[*84*] Partridge, P. G., *Acta Metallurgica,* Vol. 13, 1965, p. 517.
[*85*] Conrad, H. and Perlmutter, I., Conference Internationale sur la Métallurgie du Beryllium, Paris, Presses Universitaires de France, 1966, p. 326.
[*86*] Kocks, U. F. and Westlake, D. G., *Transactions,* The Metallurgical Society, American Institute of Mining, Metallurgical and Petroleum Engineers, Vol. 239, 1967, p. 1107.
[*87*] Dorn, J. E. and Mitchel, J. B., *High Strength Materials,* Wiley, New York and London, 1965.
[*88*] Tegart, W. J. McG., *Elements of Mechanical Metallurgy,* MacMillan, London, 1966.
[*89*] Altshuler, T. L. and Christian, J. W., *Acta Metallurgica,* Vol. 14, 1966, p. 908.
[*90*] Bolling, G. F. and Richman, R. H., *Acta Metallurgica,* Vol. 15, 1967, p. 678.
[*91*] Tenckhoff, E. in *Zirconium in Nuclear Applications, ASTM STP 551,* American Society for Testing and Materials, Philadelphia, 1974, p. 179.
[*92*] Hosford, W. F., *Metals Engineering Quarterly,* Vol. 6, 1966, p. 13.
[*93*] Rittenhouse, P. L. and Picklesimer, M. L., *Electrochemical Technology,* Vol. 4, 1966, p. 322.
[*94*] Dressler, G., Matucha, K. H., and Wincierz, P., *Canadian Metals Quarterly,* Vol. 11, 1972, p. 177.
[*95*] Conrad, H., de Meester, B., Doner, M., and Okazaki, K., Symposium on The Physics of Solid Solution Strengthening in Alloys, American Society for Metals, Chicago, Oct. 1973.
[*96*] Conrad, H., personal communication, Erlangen, 1973.
[*97*] Backofen, W. A., *Metallurgical Transactions,* Vol. 4, 1973, p. 2679.
[*98*] Tarnovskii, I. Y. et al, *Deformation of Metals During Rolling* (English ed.), Pergamon Press, New York, 1965.
[*99*] von Mises, R., *Zeitschrift angewandte Mathematik Mechanik,* Vol. 8, 1928, p. 161.
[*100*] Hill, R., *Proceedings,* Royal Society, London, Vol. A193, 1948, p. 281.
[*101*] Taylor, G. J., *Journal,* Institute of Metals, Vol. 62, 1938, p. 307.
[*102*] Bishop, J. F. W. and Hill, R., *Philosophical Magazine,* Vol. 42, 1951, pp. 414 and 1298.
[*103*] Piehler, H. R. and Backofen, W. A. in *Texturen in Forschung und Praxis,* J. Grewen and G. Wassermann, Eds., Springer, Berlin, 1969, p. 436.
[*104*] Backofen, W. A., Hosford, W. F., and Burke, J. J., *Transactions,* American Society for Metals, Vol. 55, 1962, p. 264.
[*105*] Hosford, W. F. and Backofen, W. A. in *Fundamentals of Deformation Processing,* Proceedings of the 9th Sagamore Conference, W. A. Backofen, J. J. Burke, L. F. Coffin, Jr., N. L. Reed, and V. Weiss, Eds., Syracuse University Press, New York, 1964, p. 259.

[*106*] Althoff, J., Drefahl, K., and Wincierz, P., *Zeitschrift fuer Metallkunde,* Vol. 62, 1971, p. 765.
[*107*] Chin, G. Y. and Mammel, W. L., *Metallurgical Transactions,* Vol. 1, 1970, p. 357.
[*108*] Thornburg, D. R. and Piehler, H. R., *Metallurgical Transactions,* Vol. 6A, 1975, p. 1511.
[*109*] Hosford, W. F., *Texturen in Forschung und Praxis,* J. Grewen and G. Wassermann, Eds., Springer, Berlin, 1969, p. 414.
[*110*] Dressler, G., Matucha, K.-H., and Wincierz, P. in *Proceedings,* 2nd International Conference on Structural Mechanics in Reactor Technology, Berlin, 1973, Vol. 1, C2/2.
[*111*] Ashby, M. F., *Philosophical Magazine,* Vol. 21, 1970, p. 399.
[*112*] Schulz, L. G., *Journal of Applied Physics,* Vol. 20, 1949, p. 1030.
[*113*] Chernock, W. P. and Beck, P. A., *Journal of Applied Physics,* Vol. 23, 1952, p. 341.
[*114*] Chernock, W. P., *Zeitschrift fuer Metallkunde,* Vol. 46, 1955, p. 311.
[*115*] Wilson, A. J. C., *Journal of Scientific Instruments,* Vol. 27, 1950, p. 321.
[*116*] Tenckhoff, E., *Journal of Applied Physics,* Vol. 41, 1970, p. 3944.
[*117*] Bunge, H. J., *Mathematische Methoden der Texturanalyse,* Akademie Verlag, Berlin, 1969.
[*118*] Williams, R. O., *Transactions,* The Metallurgical Society, American Institute of Mining, Metallurgical and Petroleum Engineers, Vol. 242, 1968, p. 104.
[*119*] Chin, G. Y., Mammel, W. L., and Dolan, M. T., *Transactions,* The Metallurgical Society, American Institute of Mining, Metallurgical and Petroleum Engineers, Vol. 239, 1967, p. 1854.
[*120*] Bunge, H. J., *Kristall und Technik,* Vol. 5, 1970, p. 145.
[*121*] Siemes, H., *Zeitschrift fuer Metallkunde,* Vol. 58, 1967, p. 228.
[*122*] Calnan, E. A. and Clews, C. J. B., *Philosophical Magazine,* Vol. 41, 1950, p. 1085; Vol. 42, 1951, p. 616; and Vol. 43, 1951, p. 919.
[*123*] Hobson, D. O., *Transactions,* The Metallurgical Society, American Institute of Mining, Metallurgical and Petroleum Engineers, Vol. 242, 1968, p. 1105.
[*124*] Tenckhoff, E., *Texturbildung in Zirkonium,* Vortrag auf der DGM-HVS Nürnberg, 1975.
[*125*] Tenckhoff, E., *Metallurgical Transactions,* Vol. 9A, 1978, p. 1401.
[*126*] Grewen, J., "Textures of Hexagonal Metals and Alloys and Their Influence on Industrial Application," Pont-à-Mousson Conference on Texture, 1973.
[*127*] Burgers, W. G., Fast, J. D., and Jacobs, F. M., *Zeitschrift fuer Metallkunde,* Vol. 29, 1937, p. 410.
[*128*] Nerses, V., Report NMI-1222, U.S. Atomic Energy Commission, 1960.
[*129*] Tuxworth, R. H., Report CRMET-901, Atomic Energy of Canada, Ltd., 1960.
[*130*] Laidler, J. J., Report HW-64815, U.S. Atomic Energy Commission, 1960.
[*131*] Wood, D. S., Winton, J., and Watkins, B., *Electrochemical Technology,* Vol. 4, 1966, p. 250.
[*132*] Okada, T., Hirano, H., and Kunimoto, N., *Electrochemical Technology,* Vol. 4, 1966, p. 365.
[*133*] Sturcken, L. F. and Duke, W. G., Report DP-607, U.S. Atomic Energy Commission, 1961.
[*134*] Hindle, E. D. and Slattery, G. F., *Journal,* Institute of Metals, Vol. 93, 1964–65, p. 565.
[*135*] Cheadle, B. A. and Ells, C. E., *Transactions,* The Metallurgical Society, American Institute of Mining, Metallurgical and Petroleum Engineers, Vol. 233, 1965, p. 1044.
[*136*] Cheadle, B. A., Ells, C. E., and Evans, W., *Journal of Nuclear Materials,* Vol. 23, 1967, p. 199.
[*137*] Tenckhoff, E. and Rittenhouse, P. L. in *Application Related Phenomena in Zirconium and its Alloys, ASTM STP 458,* American Society for Testing and Materials, Philadelphia, 1970, p. 50.
[*138*] Tenckhoff, E., *Zeitschrift fuer Metallkunde,* Vol. 61, 1970, p. 64.
[*139*] Tenckhoff, E. and Rittenhouse, P. L., *Zeitschrift fuer Metallkunde,* Vol. 63, 1972, p. 83.

[*140*] Picklesimer, M. L., "A Preliminary Examination of the Formation and Utilization of Texture and Anisotropy in Zircaloy-2," ORNL-TM-460, Oak Ridge National Laboratory, Oak Ridge, TN, 1963.
[*141*] Norton, J. T. and Hiller, R. E., *Transactions,* American Institute of Mining, Metallurgical and Petroleum Engineers, Vol. 99, 1932, p. 190.
[*142*] McHargue, C. J. and Hammond, J. P., *Transactions,* The Metallurgical Society, American Institute of Mining, Metallurgical and Petroleum Engineers, Vol. 197, 1953, p. 57.
[*143*] Tenckhoff, E. and Rittenhouse, P. L., *Journal of Nuclear Materials,* Vol. 35, 1970, p. 14.
[*144*] Hobson, D. O. and Rittenhouse, P. L., *Transactions,* The Metallurgical Society, American Institute of Mining, Metallurgical and Petroleum Engineers, Vol. 245, 1969, p. 797.
[*145*] Steward, K. P. and Cheadle, B. A., *Transactions,* American Institute of Mining, Metallurgical and Petroleum Engineers, Vol. 239, 1967, p. 504.
[*146*] Pickmann, D. O., *Nuclear Engineering Design,* Vol. 21, 1972, p. 212.
[*147*] Rittenhouse, P. L. in *Application Related Phenomena in Zirconium and its Alloys, ASTM STP 458,* American Society for Testing and Materials, Philadelphia, 1969, p. 241.
[*148*] Tresca, H., *Comptes Rendus Hebdomadaires des Seances de la Academie de Sciences,* Vol. 59, 1864, pp. 754 and 764.
[*149*] Grzesik, D., *Zeitschrift fuer Metallkunde,* Vol. 66, 1975, p. 187.
[*150*] Stehle, H., Steinberg, E., and Tenckhoff, E. in *Zirconium in the Nuclear Industry, ASTM STP 633,* A. L. Lowe, Jr., and G. W. Parry, Eds., American Society for Testing and Materials, Philadelphia, 1977, p. 486.
[*151*] Holicky, M. J. and Schroeder, I., *Journal of Nuclear Materials,* Vol. 44, 1972, p. 31.
[*152*] Dressler, G. and Matucha, K.-H. in *Zirconium in the Nuclear Industry, ASTM STP 633,* A. L. Lowe, Jr., and G. W. Parry, Eds., American Society for Testing and Materials, Philadelphia, 1977, p. 508.
[*153*] Duncombe, E., WAPD-TM-984, Bettis Atomic Power Laboratory, Pittsburgh, PA, 1973.
[*154*] Douglas, D. L., *The Metallurgy of Zirconium,* International Atomic Energy Agency, Vienna, 1971.
[*155*] Ardell, A. J. and Sherby, O. D., *Transactions,* American Institute of Mining, Metallurgical and Petroleum Engineers, Vol. 239, 1967, p. 1547.
[*156*] Fidleris, V. in *Application Related Phenomena in Zirconium and its Alloys, ASTM STP 458,* American Society for Testing and Materials, Philadelphia, 1969, p. 1.
[*157*] Holmes, J. J., *Journal of Nuclear Materials,* Vol. 13, 1964, p. 137.
[*158*] Berstein, J. M., *Transactions,* American Institute of Mining, Metallurgical and Petroleum Engineers, Vol. 239, 1967, p. 1518.
[*159*] Irvin, J. E., *Electrochemical Technology,* Vol. 4, 1966, p. 240.
[*160*] Azzarto, F. J., Baldwin, E. E., Wiesinger, F. W., and Lewis, D. M., *Journal of Nuclear Materials,* Vol. 30, 1969, p. 208.
[*161*] Veevers, K. and Rotsey, W. B., *Journal of Nuclear Materials,* Vol. 27, 1968, p. 108.
[*162*] Gilbert, E. R., *Journal of Nuclear Materials,* Vol. 26, 1968, p. 105.
[*163*] Ibrahim, E. E. in *Applications Related Phenomena in Zirconium and its Alloys, ASTM STP 458,* American Society for Testing and Materials, Philadelphia, 1969, p. 18.
[*164*] Howe, M. L., "The Annealing of Irradiation Damage in Zircaloy-2 and the Effect of High Temperature Radiation on the Tensile Properties of Zircaloy-2," Report 1024, Atomic Energy of Canada Ltd., 1960.
[*165*] Howe, M. L., "Radiation Damage in Zirconium, Zircaloy-2, and 410 Stainless Steel," Report 1484, Atomic Energy of Canada Ltd., 1962.
[*166*] Hesketh, R. V., *Journal of Nuclear Materials,* Vol. 26, 1968, p. 77.
[*167*] Piercy, G. R., *Journal of Nuclear Materials,* Vol. 26, 1968, p. 18.
[*168*] Nichols, F. A., *Journal of Nuclear Materials,* Vol. 30, 1969, p. 249.
[*169*] Bell, W. L. in *Zirconium in Nuclear Applications, ASTM STP 551,* American Society for Testing and Materials, Philadelphia, 1974, p. 199.